30th Annual Meeting of the North American Membrane Society (NAMS 2021)

Convergence - Connecting Membrane Science with Societal Challenges

Estes Park, Colorado, USA
28 August – 2 September 2021

ISBN: 978-1-7138-3531-8

Printed from e-media with permission by:

Curran Associates, Inc.
57 Morehouse Lane
Red Hook, NY 12571

Some format issues inherent in the e-media version may also appear in this print version.

Printed with permission by Curran Associates, Inc. (2021)

For permission requests, please contact North American Membrane Society (NAMS) at the address below.

North American Membrane Society (NAMS)
University of Arkansas
Ralph E. Martin Department of Chemical Engineering
3202 Bell Engineering Center
Fayetteville, AR 72701-1201
USA

Phone: +1 479 575-7926
Fax: +1 479 575-7926

names@uark.edu

Additional copies of this publication are available from:

Curran Associates, Inc.
57 Morehouse Lane
Red Hook, NY 12571 USA
Phone: 845-758-0400
Fax: 845-758-2633
Email: curran@proceedings.com
Web: www.proceedings.com

TABLE OF CONTENTS

1 - CONVERGENCE: PHARMACEUTICAL PROCESSING

2 - MEMBRANE FOULING I

3 - GAS SEPARATION I - NEW MATERIALS

4 - OSMOTIC PROCESSES (RO/FO/PRO) I

5 - MEMBRANE CONTACTORS/MEMBRANE DISTILLATION/PERVAPORATION I

6 - CONVERGENCE: ENVIRONMENTAL APPLICATIONS

7 - INDUSTRIAL INNOVATIONS AND ACADEMIA-INDUSTRY COLLABORATIONS

8 - GAS SEPARATION II - CARBON CAPTURE

9 - PROCESS INTENSIFICATION WITH MEMBRANES

10 - MATHEMATICS AND MACHINE LEARNING

11 - CONVERGENCE: MEDICINE AND PUBLIC HEALTH

12 - WATER INNOVATION USING MEMBRANES

13 - GAS SEPARATION III - NEW CONCEPTS

14 - ADVANCED METROLOGY

15 - MEMBRANES FOR ENERGY EFFICIENT BUILDINGS

16 - CONVERGENCE: ENERGY & CHEMICAL PROCESSING

17 - NAMS AWARD SESSION

18 - ELECTROCHEMICAL APPLICATIONS I

19 - MF/UF/NF I

20 - MEDICAL AND PHARMACEUTICAL APPLICATIONS

21 - ADVANCED MODULE/PROCESS DESIGN

22 - MEMBRANE FOULING II

23 - OSMOTIC PROCESSES (RO/FO/PRO) II

24 - INORGANIC MEMBRANES

25 - 3D PRINTED MEMBRANES

26 - ORGANIC SEPARATION (OSN)

27 - MEMBRANES FOR FOOD APPLICATIONS

28 - MEMBRANES FOR FOOD APPLICATIONS

29 - MEMBRANES FOR FOOD APPLICATIONS

30 - MEMBRANES FOR FOOD APPLICATIONS

NAMS 2021

The North American Membrane Society

Convergence

Connecting Membrane Science with Societal Challenges

30th Annual Meeting
Estes Park, Colorado
Ridgeline Hotel

Aug 28 –Sep 2, 2021

Abstract Book

Meeting Chairs:

Uwe Beuscher, W.L. Gore & Associates, Inc.

Yifu Ding, University of Colorado - Boulder

John Pellegrino, University of Colorado - Boulder

Technical Program Grid NAMS 2021

	start	end	Ballroom C/D	Ballroom B/E	Ballroom A/F	Trail Ridge	Lumpy Ridge
MONDAY Aug-30	8.00	9.00	Plenary: Peter Green – National Renewable Energy Laboratory (NREL)				
	9.30	12.30	1 Convergence: Pharmaceutical	2 Membrane Fouling I	3 Gas Separation I	4 Osmotic Processes I	5 Contactors MD/PV I
	14.00	17.00	6 Convergence: Environmental	7 Industry Session	8 Gas Separation II	9 Process Intensification	10 Mathematics and ML
	17.30	18.30				NAMS Business Meeting	
	19.00	22.00	Poster Session I (Courtyard or Foyer/Atrium; weather permitting)				
TUESDAY Aug-31	8.00	9.00	Plenary: Kristi Anseth – University of Colorado Boulder				
	9.30	12.30	11 Convergence: Medicine	12 Water Innovation	13 Gas Separation III	14 Advanced Metrology	15 Membranes for Buildings
	19.00	22.00	Poster Session II (Courtyard or Foyer/Atrium; weather permitting)				
WEDNESDAY Sep-01	8.00	9.00	Plenary: Thomas Schäfer – POLYMAT, University of the Basque Country				
	9.30	12.30	16 Convergence: Chemical	17 NAMS Awards	18 Electro-chemical I	19 MF/UF/NF I	20 Membranes for Med/Pharm
	14.00	16.00	21 Advanced Module/Process	22 Membrane Fouling II	23 Osmotic Processes II	24 Inorganic Membranes	25 3D printed membranes
	16.00	17.30	Debate: More Material Science?	Debate: Funding valley of death?	Debate: Is water public?		
	19.00	22.00	Conference Banquet (Estes Park Events Complex)				
THURSDAY Sep-02	8.00	10.00	Alan Michaels Recipient Session				
	10.30	12.30	26 Organic Separations	27 Membranes for Food	28 Electro-chemical II	29 MF/UF/NF II	30 Contactors MD/PV II

Plenary 1

Dr. Peter F. Green
National Renewable Energy Laboratory

Navigating pathways that would enable the clean energy transition

Monday, August 30, 8:00 am – 9:00 am
Estes Park Events Complex

Peter F. Green has served as deputy laboratory director for Science and Technology and chief research officer of NREL and senior vice president of the Alliance for Sustainable Energy—the company that operates NREL—since 2016. He is responsible for NREL's science and research goals, strengthening the laboratory's core capabilities, and enhancing NREL's research portfolio. He also oversees the Laboratory Directed Research and Development Program, NREL-university interactions, and the postdoctoral research program. Prior to NREL, Green spent 11 years at Sandia National Laboratories, and then became the B.F. Goodrich Endowed Professor of Materials Engineering at University of Texas. In 2005, he joined the University of Michigan as chair of the Department of Materials Science and Engineering, and the Vincent T. and Gloria M. Gorguze Endowed Professor of Engineering. He served as director of a DOE Center: Center for Solar and Thermal Energy Conversion. Green's prior leadership experience includes serving as president of the MRS and advisory boards for the national academies, national laboratories, and universities. He is the elected fellow of MRS, APS, ACeraS, RSC, and the AAAS. He holds bachelor's degrees in physics from Hunter College and a doctorate in materials science and engineering from Cornell University.

Abstract

A decarbonized energy system by 2035 and achieving net zero emissions during the next three decades pose unprecedented, yet surmountable, challenges. An essential aspect of any successful strategy is to establish a future energy system that uses power generated from diverse non-fossil sources. Such a system is envisioned to be coupled with a hydrogen infrastructure (H2@Scale), where hydrogen provides services such as supplemental power generation and seasonal energy storage for the power grid. CO2, captured from the atmosphere and from various industrial processes, together with hydrogen, would provide feedstocks for the synthesis of chemicals, fuels and materials, via various biotic and abiotic processes. The future energy system, moreover, is envisioned to include autonomous control of millions of distributed energy devices, including energy storage, electric vehicles, LEDs, electric heating and a range of smart home devices. It must be secure, reliable, resilient and support economic growth. While several studies discuss scenarios for viable pathways toward a clean energy economy, the final pathway remains uncertain. It will not be based solely on S&T advances, but also other factors, including technology adoption rates, social acceptance, economic investments, and policy. NREL, with its vision for a clean energy world, established three key objectives that would contribute toward enabling the clean energy transition: integrated energy pathways, electrons-to-molecules (power-to-X) and circular economy for energy materials. These objectives, as well as the current and future status of renewable energy technologies, including the future energy system, will be discussed.

Plenary 2

Dr. Kristi Anseth
University of Colorado Boulder

Biomaterials as Synthetic Extracellular Matrices: Designer Materials for Tissue Engineering

Tuesday, Aug 31, 8:00 am – 9:00 am
Estes Park Event Complex

Dr. Anseth is a Distinguished Professor, Tisone Professor of Chemical and Biological Engineering, and Head of Academic Leadership of the BioFrontiers Institute at the University of Colorado at Boulder. Her research interests lie at the interface between biology and engineering where she designs new biomaterials for applications in drug delivery and regenerative medicine.

Dr. Anseth is an elected member of the National Academy of Engineering (2009), the National Academy of Medicine (2009), the National Academy of Sciences (2013), and the National Academy of Inventors (2016). She is also a Fellow of the American Association for the Advancement of Science, the American Institute for Medical and Biological Engineering, American Institute of Chemical Engineers, and the Materials Research Society. Dr. Anseth currently serves on the Board of Directors of the American Institute of Chemical Engineers, Board of Trustees for the Gordon Research Conferences, on the Scientific Advisory Board of the Allen Institute, and is an editor for Biomacromolecules and Progress in Materials Science. Dr. Anseth's work has been acknowledged through many prestigious awards including Herman F. Mark Award for polymer chemistry (2021), Society for Biomaterials Founders Award (2020), and L'Oreal UNESCO for Women in Science Laureate for North America (2020), to name a few. Dr. Anseth received her B.S. from Purdue University and Ph.D from University of Colorado Boulder.

Abstract

My research group is interested in the development of polymeric biomaterials that can interface with cells and promote tissue regeneration and repair. From a fundamental perspective, we seek to decipher the critical extracellular matrix (ECM) signals that are relevant for tissue development, regeneration, and disease and then design materials that integrate these signals. From an applied perspective, we use this knowledge to engineer materials that can promote tissue regeneration in vitro and in vivo. This talk will illustrate our recent efforts towards the synthesis of new hydrogel chemistries for 4D cell culture and regenerative medicine, and how one can dynamically control biochemical and biophysical properties through orthogonal, photochemical click reaction mechanisms. Some specific examples will include the design of hydrogels that promote musculoskeletal tissue regeneration, materials-directed growth of intestinal organoids from a single stem cell, and super-swelling matrices to visualize cell-matrix interactions with unprecedented resolution. These efforts will then be placed in the context of designing precision biomaterials to address demands for patient specific products and treatments.

Plenary 3

Dr. Thomas Schafer
University of the Basque Country

Biological and Artificial Membranes: Similarity, Congruence ...Convergence?

Wednesday, Sep 1, 8:00 am – 9:00 am

Dr. Schäfer is Ikerbasque Research Professor at the Basque Centre for Macromolecular Design and Engineering (POLYMAT) in Donostia/ San Sebastián, Spain. He leads a multi-disciplinary research group in the field of nanobiotechnology/ membrane separations initially funded by the European Research Council. Thomas earned his BSc from Friedrich-Alexander University Erlangen-Nürnberg and MSc in Chemical Engineering with focus on animal cell culture from the Technical University Hamburg. He obtained his PhD from Universidade Nova de Lisboa in membrane separation processes and has worked as a postgraduate and postdoctoral researcher at Wageningen University, University of New South Wales, University of Siena, University of Pisa, as well as University of Tarragona.

Thomas has served as elected member of the council of the EMS and co-organizes the biannual "Imagine Membrane" conference series in close collaboration between NAMS and EMS. He has been very active in discussing the future of young researchers in Europe as Chair of the COST Targeted Network "Sci-Generation" and as co-founder of the Young Academy of Europe. He also is co-founder and CTO of the spin-off company "Surphase".

Abstract

Being able to mimic the function and specificity of biological membranes is frequently considered as one of the holy grails in membrane development. Similarly, a staunch and widely accepted terminology clearly distinguishes "bio"macromolecules from "common" macromolecules creating an apparent barrier between the two fields. Using desoxyribonucleic acids (DNA) as an example, in this talk I will make an attempt to decipher what is truly different in "bio", if anything at all. While DNA is mostly known for carrying genetic information, it will be shown that is an excellent building block for creating nanostructures up to membranes. I will discuss how far a biomacromolecule and respective membranes can be used beyond the canonical environment, and on what cost. Systematic development of such membranes requires adequate characterization techniques that in themselves can feed back into membrane technology. During the talk I will highlight opportunities, potential pitfalls, and what might possibly be rather "hype".

Thank you to our Sponsors

DIAMOND SPONSOR

NSF

GOLD SPONSORS

Anton Paar

AIR PRODUCTS

Membrane Solutions

NSF IUCRC • Membrane Science, Engineering & Technology Center

MAST

MilliporeSigma

Chevron

BRONZE SPONSORS

EVONIK
Leading Beyond Chemistry

Gale & DB Bhattacharyya

HIDEN ISOCHEMA

membranes
an Open Access Journal by MDPI

Georgia Tech ISSEC

GORE

MTR
MEMBRANE
TECHNOLOGY & RESEARCH

Together, improving life

1a - Novel Membrane Processes and Devices

Gastón de los Reyes (SPF Technologies)*

MF and UF membranes and their corresponding membrane devices have been in existence for at least half a century and are widely used in the manufacturing of biopharmaceuticals. In this talk I will review three different membrane-based technologies invented in the last decade that dramatically change how membranes are used, and how they fit into the new biopharm paradigm of process intensification.

Single-pass Tangential Flow Filtration (SP-TFF™) is a new TFF modality commercialized a decade ago by Pall Corporation. SP-TFF transforms the TFF process from a batch to a continuous process, reducing the residence time of the biomolecule from hours to minutes. It is now widely adopted in biopharm having found several new and high-value applications.

ChromaWeb™devices are adsorptive membrane devices in the form of stackable cassettes, superior to conventional membrane chromatography devices. Stackability enables ChromaWeb cassettes to be assembled into large arrays that can be tailored to the size of the batch. When operated under Hyper-Productive®rapid-cycling processes they become single-use devices, wherein the life of the adsorptive media is consumed in a single batch. ChromaWeb cassettes are the most practical format for process-scale membrane chromatography.

Ultra-thin Channel Filtration (UTF™) is a novel UF concentration/washing device and process that is best described as the "anti-TFF." In contrast to TFF, UTF does not try to mitigate concentration polarization; on the contrary, it exploits concentration polarization in a novel semi-continuous, rapidly-cycled process. It promises to deliver both single-pass concentration and washing of biomolecules in a manner superior to SP-TFF.

Although these innovations leverage existing membrane technologies with novel devices and/or novel processes, they will likely create opportunities for important membrane innovations.

1b - Membrane applications in integrated continuous bioprocessing (V)

Anurag S Rathore (Indian Institute of Technology Delhi)* ; Garima Thakur (Indian Institute of Technology Delhi)

Formulation of biotherapeutics using single pass tangential-flow filtration (SPTFF) is a critical step in continuous manufacturing processes for many drugs including monoclonal antibodies (mAbs). The concentration of the mAb in the final formulation is a critical quality attribute which affects safety and efficacy. The final concentration of the drug is determined by the flux across the module in a single pass, with no scope for repeated recirculation. Therefore, it is critical to design the membrane configuration and select the operating parameters such that the concentration target is met in a single pass. We develop a hybrid model for flux across a single membrane and use it as a building block to model complex SPTFF configurations and facilitate in-silico design of customized SPTFF configurations resulting in up to 40% savings in membrane area. We also propose an empirical control strategy to control the output concentration of a continuous ultrafiltration step. The control strategy leverages in-line concentration, flowrate and pressure sensors, including near infrared spectroscopy (NIRS) flow cells to measure the concentration of mAb in the feed and retentate streams in the range 0.5-200 g/L. The proposed system is the first approach that consistently achieve concentration targets over long continuous campaigns in spite of feed variability.

1c - Virus Retentive Filters in Biotech Viral Safety: 2021 Update (V)

Kurt Brorson (Parexel International)*

Virus retentive filtration is an evolving technology; new, more robust filter types have been introduced, along with the concept of use of virus filters as a barrier prior to cell culture. Related guidance is keeping up with these changes; the upcoming ICH Q5A revision is expected to address on-going regulatory issues for virus filters. In addition, the Parental Drug Association has promulgated "ANSI/PDA Standard 04-2021 Phage Retention Nomenclature Rating for Small and Large Virus-Retentive Filters", an ANSI accredited system for virus filter nomenclature based on bacteriophage retention. The same organization is near finalization of an update to their Technical Report 41, "Virus Retentive Filters", which has served as a benchmark and educational guide for end users since 2005. In this talk I will provide an update on this activity.

1d - Clarification strategies for the adeno-associated viral vector cell culture harvest: Challenges and Solutions (V)

Xiaotong Fu (Biogen)* ; John Pieracci (Biogen)

As the gene therapy field advances, hundreds of clinical trials using Adeno-associated viral vectors (AAV) are currently being investigated to identify potential gene therapy drug candidates. For those targeting the indications with a large patient population, huge demands of viral vector manufacturing capacity are expected. As a result, AAV production at larger scales, along with increased cell mass, is being developed to meet productivity demands. However, this can lead to significant downstream clarification challenges due to inefficient removal of cells/ cell debris or non-specific product binding. In this project, we evaluated different harvesting methods, combined with a variety of clarification strategies, to filtrate cell culture lysate from AAV productions. Challenges such as low step yield, or poor filtration efficiency were addressed. Potential optimizations to improve clarification performance were proposed and evaluated.

1e - Downstream purification of virus-based therapeutics - the final frontier for membrane processes? (V)

David Latulippe (McMaster University)*

Viruses are an important new class of bio-therapeutics - while certain viruses have been engineered to deliver therapeutic genetic material to specific cells, tissues, or organs, other viruses have been engineered to selectively replicate in and destroy cancer cells. For example, a 2018 paper related to a phase I trial reported very encouraging clinical responses following the injection of an oncolytic adenovirus into patients with recurrent brain tumours [1]. However, the advancement of these promising new therapies is severely limited by their exorbitant price. For example, Zolgensma, an adeno-associated virus-based treatment for pediatric patients with spinal muscular atrophy that was approved by the FDA in 2019 as a 'breakthrough therapy', has a reported cost of $2.1 million. It is well known that the process of making and purifying the virus itself is the major factor in the cost of these therapies – while there have been a number of recent advances in upstream manufacturing processes (e.g. single use bioreactors, viral sensitizers), the current downstream purification processes are still in their "infancy" with poor overall yields and thus would greatly benefit from significant improvements [2]. In this talk, I will review the various roles of membrane processes in manufacturing of virus-based therapeutics (e.g. sterile filtration using 0.22 µm-rated membranes), identify the ongoing challenges in the field (e.g. high-resolution separation of product-related impurities), and discuss some of the bold technological advancements that must be achieved to advance virus manufacturing to a realm where no process has gone before.

[1] Lang et al., J Clin Oncol. 2018 May 10; 36(14): 1419–1427

[2] National Institute for Innovation in Manufacturing Biopharmaceuticals (NIIMBL) Gene Therapy Roadmap

1f - Reducing the Bioprocessing Environmental Footprint and Saving Lives via Advanced Engineering Principles (V)

David Roush (Merck & Co., Inc.)* ; Peter Liu (Merck & Co., Inc.); Ehsan Allah Espah Borujeni (Bristol Myers Squibb); Lara Fernandez Cerezo (Merck & Co., Inc.); Mark Brower (Merck & Co., Inc.); William Rayfield (Merck & Co., Inc.)

The combination of reducing the environmental footprint while producing biologics saves lives directly and indirectly (reduced CO2 emissions, less H2O and waste). Bioprocessing has a process mass intensity index (PMI) ≥ 10x higher than small molecule API. A solution is process intensification via upstream (intensified cultures at high cell densities and product concentrations) and downstream (high capacity resins, reduced pool volumes/filter area via increased membrane loadings, SPTFF) technologies. A strategy to increase productivity is intensified fed-batch (i.e. high cell density perfusion in seed and production reactors), often accompanied with declining viabilities creating a challenge for harvest, which relies on depth filtration (DF), either alone or with disk-stack centrifugation in traditional facilities or TFF in single-use facilities, pushing to the technology to the theoretical limit. New DF address these challenges via increased solids loading capacity, reduced plastic and water usage and minimized cell lysis. It is challenging to develop/model DF for primary recovery, owing to the multiple complex interactions including adsorption (ex. proteins, nucleic acids and lipids), media variability, many solids removal/fouling mechanisms, and the dynamic nature of the feed, and the lack of a scale-up/scale-down model owing to design differences at scale. Continuous processing can potentially reduce the PMI by more than 20%. A primary recovery challenge is multiple potential fouling mechanisms from protracted processing times (up to 60 days) resulting in reduced product sieving/recovery and TMP increase. This paper will explore how engineering fundamentals (empiricism and mathematical modeling with CFD) was used to address the primary recovery challenges for intensified fed-batch via a two stage DF system, optimal design of a MF for cell retention and creative experimental designs to address operation of viral filters in continuous processing applications.

2a - Click Crosslinked Self-Assembled Zwitterionic Nanofiltration Membrane for High Salt Selectivity and Fouling Resistance

Abhishek N Mondal (Tufts university)* ; Ayse Asatekin (Tufts)

Membrane-based separations are critical for chemical separation processes. The energy efficiency, broad scalability, and portability of membrane systems make them promising for treating challenging wastewater streams. Random Zwitterionic Copolymers (R-ZACs) are unique class of linear random/statistical copolymers that combine zwitterionic and hydrophobic segments. Asatekin group previously showed that R-ZAC based membranes exhibit high flux, high selectivity with ~1.2 nm effective pore size, and unprecedented fouling resistance to a broad range of foulants. However, the pore size was difficult to adjust to address varying separations. While cross-linkable R-ZACs were developed to decrease effective pore size down to <1 nm, this required >20 minute UV treatment times, limiting scalability to roll-to-roll systems. In this present work, we have demonstrated that Thiol-Ene click chemistry for crosslinking of R-ZACs can be effective for the fabrication of robust membranes that exhibit high flux, divalent ion rejection, and fouling resistance while decreasing UV exposure time. Crosslinking lead to an enhancement of the rejection of both organic uncharged solutes and divalent salts, while retaining comparatively low NaCl rejection. For instance, a Thiol-Ene-crosslinked R-ZAC membrane showed vitamin B12 and riboflavin rejections of 99.8% and 78.7%, respectively, compared to its uncrosslinked analogue (VB12 and riboflavin rejection of 87.5% and 18%). Thiol-ene crosslinked R-ZAC membranes also showed high divalent salt rejection (e.g. 98% Na_2SO_4 rejection) whereas the monovalent salt rejection was fairly low (e.g. 32% NaCl rejection). When compared with commercial nanofiltration membrane NP-30, our membrane showed incredible fouling resistance with no irreversible flux loss. The effective pore size of these crosslinked membranes can be easily tuned by simply changing the crosslinking time; hence the selectivity of these membranes is easily tunable for achieving desired goal.

2b - Hydrogel coated MOF interlayer on polyacrylonitrile membrane exhibits excellent fouling-resistance and separation performance for produced water treatment

Akshay Modi (Ben Gurion University of the Negev)*; Zhongyi Jiang (Tianjin University); Ron Kasher (Ben Gurion University of the Negev)

Produced water (PW) from the oil and gas industry is generated in an enormous amount to meet the ever-increasing demand of the growing human population in different societal sectors. PW requires treatment before discharge to minimize environmental damage or treatment for freshwater reclamation in water stress regions. Membrane technology is used for PW treatment. However, its performance is limited by membrane fouling. Therefore, to mitigate membrane fouling, surface modification by hydrogel coated nanoparticles interlayer is one of the preferred methods, among several other methods, due to the ease in processing and negligible environmental impact. In this study, an interlayer of metal organic framework (MOF) nanoparticles was in-situ grown on a polyacrylonitrile (P) membrane, followed by coating with a grafted methacrylate hydrogel. The novelty of this study is to hydrostabilize the MOF layer on the membrane surface during long-term filtration experiments. Compared to the pristine membrane, the prepared g-MOF-P membrane showed improved hydrophilic nature, negative surface charge in the operating pH, and minimal surface roughness. The pure water permeance of the g-MOF-PAN membrane was measured to be ~13% higher (447.9 ± 6.7 L·m–2·h–1·bar–1 vs. 396 ± 6.5 L·m–2·h–1·bar–1) than that of the pristine membrane. Interestingly, the functionalized membrane (g-MOF-PAN) exhibited exceptionally high antifouling property compared to the pristine membrane (flux recovery was 98.5 ± 0.6% vs. 8.1 ± 0.7%) when tested with the oilfield produced water for three 3 h long cycles of PW filtration. Moreover, >99% oil rejection efficiency was also shown by the functional membrane, indicating its outstanding PW separation performance. Overall, the results indicated that the hydrogel coated MOF interlayer on polyacrylonitrile membrane exhibits high flux, excellent fouling-resistance, and oil rejection efficiency for produced water treatment.

2c - Micro to macro: Connecting foulant structure and mechanics with hydraulic resistance

Jose Agustin Epstein (Technion - Israel Institute of Technology); Guy Z Ramon (Technion - Israel Institute of Technology)*

Fouling remains a major operational concern in membrane-based water purification and desalination. Understanding how the pressure-driven permeate flux affects the mechanical and hydraulic properties of a foulant layer can provide important insight leading, for example, to more efficient mechanical cleaning methods. Here, we present a new approach that enables in-situ analysis of a foulant layer at the membrane surface. Specifically, we aim to connect the foulant's mechanical properties on the microscale with their macroscopic effect - the observed hydrodynamic resistance - all under pressure and with the ability to control and vary the permeate flux. A video microscopy-based particle-tracking technique is employed to visualize a foulant layer located on a membrane surface, within a custom-made filtration cell, enabling high-resolution imaging with a confocal microscope. Specifically, the motion of fluorescent tracers embedded in the foulant layer is monitored. Through the analysis of particle trajectory statistics, the mechanical properties of the foulant layer are characterized in terms of the frequency-dependent elastic and viscous components. The method was employed to study an alginate layer deposited on a Nanofiltration membrane, under a step-wise increase of the permeate flux. In each experiment, the foulant layer was visualized at different distances from the membrane surface. Our results reveal a positive correlation between the foulant's complex viscosity and its hydrodynamic resistance, in response to changes in the induced permeate flux, indicative of possible structural changes due to compaction. Further, the foulant layer exhibited depth-heterogeneity, with clear variations observed at different distances from the membrane surface, becoming more 'stiff' in adjacent to the membrane surface. The presented approach opens new possibilities for studying fouling, particularly soft, gel-like deposits typical of organic and bio-fouling.

2d - Comparison of calcium scaling in direct contact membrane distillation (DCMD) and nanofiltration (NF)

Zhewei Zhang (University of Pittsburgh)* ; Shardul Wadekar (Intel Corporation); Omkar Lokare (Gradiant Corporation); Radisav Vidic (Pitt)

Calcium scaling is commonly encountered in membrane separation processes with a profound impact on permeate flux and energy efficiency of these processes. Direct contact membrane distillation (DCMD) and nanofiltration (NF) were selected to study calcium scaling impact on thermal and pressure driven membrane processes. Calcium sulfate (CaSO4) and calcium carbonate (CaCO3) solutions with salinity in the range from 3,000 to 30,000 mg/L were used as the feed. All DCMD and NF tests were performed at identical feed temperature (i.e., 40 °C) and shear conditions (i.e., Re = 771±28). The average size of CaSO4 and CaCO3 crystals formed in the DCMD system was larger and the scale was more porous than in the NF system. The permeate flux in NF system was impacted by both calcium scales regardless of the feed salinity, while only CaSO4 scale formed at low salinity affected the performance of DCMD system. Surface crystallization in NF systems is likely extended into the membrane pores, which lead to permanent reduction in permeability. This study demonstrated that the impact of calcium scaling is affected by both feed salinity and separation driving force, and is much less severe in thermal than pressure driven membrane processes.

2e - Fouling resistant and tunable polyampholyte selective layers for salts and small organic molecules separations

Luca Mazzaferro (Tufts University)* ; Ayse Asatekin (Tufts)

Membrane separations are energy-efficient, simple, and scalable. Yet, their broader use is limited by the separation capabilities of membranes prepared by conventional methods, typically susceptible to fouling and with poor selectivity due to high pore size polydispersity. Self-assembly of functional polymeric materials is a powerful platform for designing membranes capable of new and more efficient separations. Previous researchers have developed a method for preparing such membranes by the spontaneous self-assembly of random amphiphilic copolymers, it provides a combination of scalability and good performance. Here we present a new class of self-assembled random amphiphilic copolymers with tunable selectivity and good fouling resistance based on polyampholyte random amphiphilic copolymers. Polyampholytes are copolymers with positively and negatively charged monomer subunits. In this study we demonstrate that polyampholyte random amphiphilic copolymers are highly fouling resistant, and that it is possible to tune salts and small organic molecules selectivities by controlling the ratio of positive and negative monomer subunits of our copolymers. We were able to vary the rejection of sodium sulfate from 5% to 93% with no significant change in pore size or fouling resistance. We also demonstrate that polyampholyte selective layers are a promising platform to create membranes with good performance and easy fabrication.

2f - Exploring monoclonal antibody (mAb) filtration through virus retentive membranes

Matthew W Billups (Pennsylvania State University)* ; Swarnim Ranjan (Bristol-Myers Squibb); Melissa Holstein (Bristol-Myers Squibb); Sanchayita Ghose (Bristol-Myers Squibb); Andrew Zydney (Penn State University)

The manufacture of biopharmaceuticals requires strict regulatory steps to ensure the safety of the final product. Virus filtration (VF) is a critical component of the overall viral clearance strategy. VF is typically run in normal-flow filtration with the virus captured by the small (approximately 20 nm) pores in the membrane while the smaller biopharmaceutical is collected in the filtrate. One of the critical issues in the design of VF processes is membrane fouling, which causes a decrease in flux during constant pressure operation. Recent data obtained with different monoclonal antibody (mAb) products shows dramatic differences in fouling. The objective of this work is to develop a more fundamental understanding of the key factors leading to the different fouling behavior of several mAb products.

Experiments were performed using constant pressure normal flow filtration at 30 psi with Viresolve Pro virus filters. In each case, there was an immediate drop in filtrate flux (compared to the corresponding buffer flux). The "fouling" was nearly completely reversible; flushing the membrane with buffer after the mAb filtration resulted in almost complete recovery of the original buffer flux. The different mAbs showed different initial flux decline, ranging from <20% at a 20 g/L concentration to more than 90% at a concentration of only 4 g/L.

Additional insights into the fouling behavior were obtained using analytical hydrophobic interaction chromatography (HIC), which provides a measure of the relative hydrophobicity of the mAb based on its interaction with a hydrophobic surface. The mAb that displayed the greatest degree of fouling was the most hydrophobic, suggesting that the increase in fouling were governed by intermolecular hydrophobic interactions. These results provide important insights into the fouling mechanisms during VF as well as an initial framework for the development of appropriate strategies to mitigate fouling based on the specific properties of the mAb.

3a - Rational design of highly selective and plasticization resistant PIMs inspired by competitive sorption

Katherine Mizrahi Rodriguez (MIT)* ; Francesco Maria Benedetti (Massachusetts Institute of Technology); Naksha Roy (MIT); Albert Wu (MIT); Zachary Smith (MIT)

Polymers of intrinsic microporosity (PIMs) have shown excellent pure-gas separation performance due to their inefficient packing and high free volume. Their out-of-equilibrium packing morphologies, however, make PIMs susceptible to physical aging and their intrachain rigidity alone has shown insufficient to mitigate plasticization. A number of studies on the mixed-gas transport of PIMs have shown the effects of competitive sorption on performance, where gases with high polymer affinity (e.g., CO_2) can reduce the sorption of co-penetrants in a mixture (e.g., CH_4, N_2) and increase sorption selectivity. However, due to CO_2-induced plasticization at high pressures, decreases in diffusion selectivity can outweigh beneficial competition effects. This trade-off in performance is especially detrimental for PIMs with little CO_2 affinity and poor plasticization resistance, as they rely on diffusion-selective transport that may be reduced at high pressure.

Here, we report on mixed-gas and high-pressure transport properties of six functionalized PIMs. Low-pressure mixed-gas tests indicate a relationship between CO_2 sorption affinity and enhancements in CO_2/CH_4 mixed-gas selectivity compared to pure-gas for all PIMs considered. The best results are reported for amine-functionalized PIM-1 (PIM-NH_2), which shows an unprecedented 2.4- and 3.5-fold increase in mixed-gas CO_2/CH_4 and CO_2/N_2 mixed-gas selectivity, respectively, compared to pure-gas. PIM-NH_2 films retain high mixed-gas selectivity up to a total mixed-gas pressure of 26 bar. Pure-gas sorption and mixed-gas permeation performance for the six PIMs were compared across a range of reported microporous polymers, elucidating structure/property relationships that can enable rational design of polymers capable to perform well in industrially relevant scenarios. Results demonstrate the promise of primary amine functionalization for developing highly sorption-selective and plasticization-resistant membranes for gas separations.

3b - Intrinsically Microporous Ladder Polymers Containing Pentiptycene Frameworks Exhibiting Physical Aging-Enhanced Gas Separation Performance

Tanner Corrado (University of Notre Dame)* ; Ruilan Guo (University of Notre Dame)

Gas separation membrane technology has seen major advancement through the study of high-free-volume polymers such as PIM-1. Unfortunately, this free volume is largely conformation-based, with polymers often experiencing significant declines in permeability from free volume collapse during physical aging. In this work, we exploit pentiptycene's versatile architecture as a building block to deliver microporous ladder polymers containing configuration-based free volume that is tunable and non-collapsible. Copolymers with PIM-1 were created utilizing the pentiptycene framework in a variety of monomer configurations (S- or C-shaped) containing tailorable substituent groups (n- or iso-propoxy units) to explore the effect of pentiptycene backbone configuration and degree of pentiptycene incorporation on gas transport properties. Copolymers incorporating 17 mol % of either the S- or C-shape unit with the isopropoxy substituent saw H_2 permeability increases of 17-19% and 21-31% increases in H_2/CH_4 selectivity relative to PIM-1, placing them well above the 2008 upper bound. Manipulation of substituent groups on pentiptycene, i.e., switching branched isopropoxy to linear n-propoxy, on the S-shape copolymer yielded O_2/N_2 separation performance well above the 2008 upper bound. Increasing pentiptycene content from 17 to 25 mol% in the S-shape isopropoxy substituted copolymer led to near vertical movement toward the 2015 upper bound due to the intrinsic configurational free volume of pentiptycene moieties. More importantly, these pentiptycene-based PIM-like ladder polymers showed atypical aging profiles, exhibiting aging-enhanced permeability accompanied by maintained selectivities, leading to aged film performance exceeding the 2015 upper bound for O_2/N_2, in sharp contrast with typical PIM-like ladder polymers. In this talk, we will explore synthesis and characterization of the series, with discussions focused on elucidation of the fundamental structure-property relationships.

3c - Enhanced CO_2/CH_4 separation performance of polymerized ionic liquid–ionic liquid membranes made with multifunctional ionic cross-linkers

Chamaal Karunaweera (University of Colorado Boulder)* ; Bénédicte Poulain (Total); Richard Noble (Deparment of Chemical and Biological Engineering, University of Colorado - Boulder); Douglas Gin (Department of Chemistry, University of Colorado - Boulder)

Mixed-matrix membranes (MMMs) comprised of a cross-linked polymerized ionic liquid (PIL) matrix, an ionic liquid (IL) interfacial agent, and microporous charged zeolite particles exhibit enhanced CO_2/CH_4 separation performance. However, it is difficult to form defect-free, thin-film composite (TFC) membranes of these MMMs because of heterogeneous nature of their starting mixtures. Composite membranes comprised of just a PIL and an IL can be more easily scaled-up in a defect-free manner, but cross-linked PIL-IL membranes generally show poorer gas separation performance compared to PIL-IL-zeolite MMMs. In this presentation, we show that the CO_2 permeability and CO_2/CH_4 selectivity of PIL-IL membranes can be improved by using novel cross-linkers instead of traditional uncharged hydrocarbon cross-linkers. Also, by changing the ratio of PIL to IL, the performance of these membranes can be further enhanced. For instance, a PIL-IL membrane fabricated with a new cross-linker exhibited a ca. 44% increase in CO_2 permeability and a ca. 12% increase in CO_2/CH_4 selectivity compared to an analogous PIL-IL-SAPO-34 MMM. All membranes were tested using a CO_2:CH_4 (50:50) feed mixture at elevated pressures (20–40 bar) and temperature (50 oC) to simulate real-world operating conditions for CO_2/CH_4 separations.

3d - Analysis of the Transport of Guest Molecules in Molecularly Mixed Composite Membranes Containing Porous Organic Cages

Matthew Rivera (Georgia Institute of Technology)* ; Ryan Lively (Georgia Tech)

Individual molecules with intrinsic porosity, such as porous organic cages (POCs), have significant potential to improve a wide range of separations applications. An exemplar application is the blending of POCs with polymers to make molecularly mixed composite membranes (MMCMs). The molecular-level interaction between individual cage molecules and polymer chains results in a "solid-solution" that avoids longstanding interfacial issues associated with mixed matrix membranes. Moreover, as the cages are soluble in polymer solutions, the processing of these composites can be easily adapted to established polymer-based technologies. While their potential to improve membrane applications is exciting, MMCMs are still a relatively new development, and underlying transport processes within the membrane are not well understood. Here, we offer a detailed interpretation of guest transport through these solid solutions. We demonstrate how the presence of cage molecules affects polymer chain motions that can impact guest transport through the polymer phase. We also show how cage loading affects membrane free volume. We find that permeation increases steadily up to an inflection point then increases much more rapidly. We hypothesize this is due to the creation of additional free volume in the bulk polymer. This work provides preliminary information on the nature of cage-polymer interactions and guest transport in MMCMs to aid their future adaptation to industrially-relevant separation units.

3e - High-temperature hydrogen/hydrocarbon separations in asymmetric carbon molecular sieve hollow fiber membranes

Lu Liu (University of Maryland)* ; Dongxia Liu (University of Maryland); Chen Zhang (University of Maryland)

High-temperature catalytic membrane reactors are rarely practiced at large scale due to the lack of scalable high-temperature separation membranes. Carbon molecular sieve (CMS) membranes are derived from carbonization of polymer precursors and can provide attractive separation performance. Due to excellent thermal resistance and chemical stability, CMS membranes are promising candidates for high-temperature membrane reactor applications. Notably, propane dehydrogenation (PDH) membrane reactors are of great interest for on-purpose propylene production. However, CMS membranes have not been studied for PDH membrane reactors, which require permeation temperature above 500 oC.

In this talk, we will present high-temperature hydrogen/propane separation performance of asymmetric polyimide-derived CMS hollow fiber membranes in non-oxidative PDH conditions. The roles of membrane pyrolysis condition, permeation temperature, and feed composition on high-temperature hydrogen/propane separation performance will be presented. The effects of high-temperature hydrogen and propane exposure on CMS pore structure and transport properties will be discussed. Long-term stability of the CMS hollow fiber membranes under high-temperature permeation will also be shown.

3f - Fluorinated Vinyl-addition Polynorbornene for Natural Gas Separation

Xinyi Wang (University of Tennessee, Knoxville)* ; Trevor Wilson (University of Tennessee, Knoxville); Chris Maroon (University of Tennessee, Knoxville); Brian Long (University of Tennessee)

Natural gas is important due to its high burning efficiency in processes such as electrical energy generation. However, crude natural gas contains two contaminants, CO_2 and H_2S, which must be eliminated before transportation. Two primary techniques to accomplish this task are amine absorption and polymeric membrane-based separations, though the latter has received increasing attention as a more energy-efficient passive process. Recently, our group found that a series of alkoxysilyl substituted vinyl-addition polynorbornenes (VAPNBs) show exceptional H_2S/CH_4 separation performance, as well as good separation capabilities for CO_2/CH_4. Inspired by prior literature in which introducing fluorinated moieties has been shown to improve CH_4 separation performance, we hypothesized that similar advances may be observed upon fluorination of our VAPNBs.

Herein, we describe our efforts to develop a series of fluorinated VAPNBs for natural gas separation. These polymers were synthesized using common vinyl-addition polymerization catalysts, achieving high molecular weight and forming robust free-standing films. As an example of their gas separation performance, fluorinated polymers such as trifluoroethoxysilyl-substituted VAPNB displayed a CO_2 permeability of 920 Barrer with CO_2/CH_4 selectivity of 9.50. These values represent a 67% increase in CO_2/CH_4 selectivity as compared to its non-fluorinated VAPNB analog, whereas CO_2 permeability only decreased by 10%, thereby leading the performance closer to the theoretical upper bound for this separation. Sorption studies were also performed to better understand the fundamental rationale for this improved performance. We found that diffusion selectivity decreased as a function of increasing fluorine content, whereas sorption selectivity increased sharply. To our surprise, this increase in sorption selectivity arises primarily due to increased CO_2 sorption within the fluorinated materials, while CH_4 sorption was relatively unchanged.

4a - Local density inhomogeneities govern transport properties in reverse osmosis membranes

Michael Geitner (The Pennsylvania State University)* ; Tyler E Culp (Pennsylvania State University); Abhishek Roy (Dow); Mou Paul (Dow); Steve Jons (DuPont); Jeffrey D Wilbur (DuPont Water Solutions); Manish Kumar (University of Texas at Austin); Enrique G

A mechanistic description of the effect of microstructure on membrane properties will accelerate the development of next-generation materials for water separation. Here, we characterize the internal morphology of the polyamide active layers of reverse osmosis membranes via scanning transmission electron tomography. This technique allows 3D reconstructions of the active layer to be obtained with nanometer resolution, yielding how the void fraction, surface area, and local density vary throughout the film. Fluid-flow simulations through our 3D reconstructions accurately predicts the water flux, and reveals that controlling the mass distribution at the nanometer scale is key to maximize water production. We also examine membranes challenged with nanoparticles, to reveal locations for solute rejection and pathways for flow. Overall, our work represents a new approach to characterize membranes used for water purification, and points to a new design criteria for water filtration membranes, the control of density inhomogeneities.

4b - Quantifying uncertainties in water-solute selectivity of reverse osmosis membranes caused by not accounting for concentration polarization

Mikayla D Armstrong (University of North Carolina at Chapel Hill)* ; Orlando Coronell (University of North Carolina at Chapel Hill)

A guiding goal in reverse osmosis technology is to make a membrane that breaks through the trade-off frontier between water-solute selectivity (A/B, bar-1) and water permeance (A, Lm-2.h-1.bar-1), where selectivity is defined as the ratio between A and solute permeance (B, Lm-2.h-1). Plots of A/B versus A are commonly used to compare trends of membranes within a study and more broadly with the literature. There are three ways to calculate B: accounting for, estimating, and neglecting concentration polarization. Accounting for concentration polarization is resource-intensive, so many studies either estimate or neglect it. However, not accounting for concentration polarization can result in underestimations of A/B (i.e., overestimations of B). The corresponding unreported (and in many cases unknown) uncertainties in A/B values could represent an impediment to unequivocally identifying manufacturing approaches that break through the trade-off frontier; however, we did not find a study in the literature quantifying the uncertainty in A/B values associated with the different B quantification approaches. Accordingly, this study aims to determine whether the selection of a solute permeance (B) quantification approach affects conclusions drawn on membrane performance or trends concerning the trade-off frontier. Specific objectives are to (1) compare B and A/B values obtained across the three B quantification approaches and (2) evaluate whether differences across the three approaches depend on membrane flux, applied pressure, or solute. Initial results show that neglecting concentration polarization resulted in overestimations of solute permeance ranging from 2 to 178%. A high flux membrane had higher overestimations (3-4x) than a low flux membrane, and overestimations increased with pressure. Our work contributes to a better understanding of the effects of assumptions about concentration polarization on the calculation of solute permeance and water-solute selectivity.

4c - Molecular Layer Deposition for the Fabrication of Desalination Membranes with Tunable Metrics

Brian Welch (University of Colorado)* ; Olivia McIntee (University of Colorado Boulder); Tyler Myers (University of Colorado); Alan R. Greenberg (Membrane Science, Engineering, and Technology (MAST) Center, Department of Mechanical Engineering, University

The recent advancement of semiconductor devices to the near-atomic scale necessitated the development of atomic layer processing methods, including molecular layer deposition (MLD). This gas-phase deposition technique creates semipermeable polymer films with precise control of composition and thickness. This work describes the use of MLD to produce thin-film composite reverse osmosis membranes. Aromatic polyamide films as thin as 0.5 nm were applied to NF270 nanofiltration membranes with sequential dosing of vapor phase m-phenylenediamine and trimesoyl chloride. Within a few molecular layers, desalination performance was affected. As film thickness increased to 15 nm, performance progressed from nanofiltration to reverse osmosis metrics in terms of salt rejection and water permeance. With film thickness >5 nm, rejection values exceeded those measured for a small sample of commercial membranes. Increased exposure time to MLD reactants led to increased rejection values, suggesting infiltration of the reactants into the existing polyamide structure. In all cases, a tradeoff between rejection and permeance was observed. Atomic force microscopy measurements indicate that MLD enhancement led to removal of small-scale roughness features and resulted in a root mean square roughness difference of <0.1 nm from the substrate. These initial MLD studies represent a novel processing approach that offers a potential pathway for the fabrication membranes with finely tailored properties.

4d - Molecular Methods for Assessing the Morphology, Topology, and Performance of Polyamide Membranes

Riley Vickers (University of North Carolina at Chapel Hill)* ; Timothy Weigand (University of North Carolina at Chapel Hill); Casey Miller (University of North Carolina at Chapel Hill); Orlando Coronell (University of North Carolina at Chapel Hill)

Atomistic molecular dynamics (MD) simulations are a promising pathway to determine fundamental property-performance relationships of reverse osmosis active layers. All permeating molecules must travel through the active layer free volume. MD simulations can be utilized to access the spatial and temporal scales required to quantify the interactions between the active layer and permeating molecules. MD simulations of polyamide active layers utilize polymerization and hydration methods designed to mimic real-world conditions while managing the computational cost of simulation. However, no comparative analysis exists to determine the most computationally cost-effective and realistic simulation methods. Further, no tools exist to quantify the simulated active layer morphology such that active layer properties can be fundamentally related to simulated performance. Accordingly, the goal of this work is to improve aspects of MD simulation and molecular-scale analysis of polyamide active layers to advance the fundamental understanding of polyamide property-performance relationships. The objectives are: (1) to compare polymerization simulation techniques; (2) to evaluate simulated active layer hydration approaches for reducing computational cost; and (3) to advance methods for evaluating active layer structure. Initial results show that simulated active layers polymerized with a hexane solvent are not significantly different from those polymerized without hexane. Hydration simulations indicate that insertion of water molecules into aggregate pores (radius=3.5-4.5 A), without insertion into network pores (radius=2.1-2.4 A), reduces computational costs and preserves structures. Finally, we quantify polyamide structure by determining the connectivity of polyamide free volume, then evaluate path size via a cavity energetic sizing algorithm. This work highlights methods that advance realistic and efficient MD simulation of polyamide synthesis, structure, and performance.

4e - Salt and Water Transport in Reverse Osmosis: Beyond the Solution-Diffusion Model

Li Wang (Yale University)* ; Menachem Elimelech (Yale University); Yuhao Du (Yale University)

Understanding salt-water separation is critical for reverse osmosis (RO) development and optimization. The conventional solution diffusion model neglects the interactions between the different species along the transport pathway through the membrane. We develop an ion transport model considering the separation mechanisms of partitioning and the interactions between the species. Partitioning is collectively determined by Donnan repulsion, dielectric exclusion, and size exclusion. The transport through the membrane is described by the extended Nernst-Planck equation and hinderance factors. The model is validated with experimental results in terms of salt rejection and permeate flux. We then investigate the effects of hydraulic pressure and feed concentration on the water and salt permeability coefficients, and demonstrate a tradeoff between the two coefficients. We further theoretically study the dependence of salt rejection on membrane properties such as dielectric constant, pore size, and volumetric charge density. The significance of dielectric exclusion in influencing the salt rejection is highlighted. We also develop a framework to analyze the pressure drops across the membrane on the basis of the chemical potential, and identify the dominant pressure drops in transport across the membrane.

4f - Re-thinking polyamide thin film formation: how interfacial destabilization dictates film morphology

Ines Nulens (KU Leuven); Ivo Vankelecom (KU Leuven); Guy Z Ramon (Technion - Israel Institute of Technology)* ; Adi Ben Zvi (Technion - Israel Institute of Technology)

Thin-film polyamide (PA) selective layers fabricated via interfacial polymerization (IP), form the core of membrane-based water purification. Despite widespread commercial application, the link between synthesis conditions and resultant film properties is not fully understood. Unravelling this fundamental knowledge is essential for designing better membranes for different applications. Much experimental work has been reported, where synthesis conditions were varied, and the resultant performance and morphology discussed. However, contradicting trends are often observed, primarily due to lack of a rigorous theoretical premise guiding the experiments. Here, we outline an approach that considers the interfacial stability of a system undergoing IP, with the notion that a transition from stable to unstable state corresponds to a morphological shift of the resultant film, from smooth to crumpled. Within the proposed framework, we re-examine data reported in the literature, encompassing a wide variety of synthesis conditions. Specifically, we identify and classify physical parameters that affect either the polymer formation rate and/or the system's susceptibility to instability. A conceptual 'phase diagram' maps morphological 'regimes' with respect to the relevant parameter space. Our analysis illustrates how PA morphology, observed by SEM images collected from the literature, transitions from smooth to crumpled, showing remarkable correlation with the defined parameter space. Re-casting results in this manner highlights not only the underlying physics, but also potential pitfalls where variations of synthesis conditions result in conflicting effects, making it difficult to define clear trends. New experimental data, cast within the developed physical framework and supplemented by theoretical quantification, will strengthen the fundamental understanding of the process for a given chemistry, with the goal of providing a clear physical toolbox for IP-based membrane design.

5a - Wetting in membrane distillation: modes, mechanisms, and metrics

Allyson L McGaughey (University of Southern California)* ; Shounak Joshi (University of Southern California); Amy Childress (University of Southern California)

Membrane distillation (MD) is a separation process in which the temperature difference between a warmer, saline feed and cooler, pure distillate stream drives water flux through a hydrophobic, unwetted membrane. The membrane supports a vapor gap that separates the streams; if pore wetting occurs, the streams freely mix, resulting in contamination of the distillate stream. Therefore, membrane wetting resistance is critical.

Membrane wetting resistance is often characterized by liquid entry pressure, which depends on liquid-vapor surface tension (a feed solution property), pore size (a membrane property), and contact angle (a property of the membrane and the feed solution). Typically, the measured contact angle or the intrinsic contact angle (i.e., contact angle of the smooth, solid membrane material) are used to characterize wetting resistance. In a previous study, we observed that internal or pore wall hydrophobicity may contribute to wetting resistance, especially when partial pore wetting occurs. We recently developed a method to characterize, for the first time, the internal contact angle of hydrophobic PTFE membranes based on an extended Washburn method originally developed for fuel cell materials; results will be presented.

We also develop a solute mass-balance approach to quantify liquid and vapor flux in MD and we use the approach to define two modes of wetting: constant and increasing. Results enable us to link wetting mechanisms with wetting modes. We also build on recent research that introduced a fundamental wetting resistance/vapor flux trade-off in MD due to partial wetting of membranes with lower hydrophobicity. Here, we quantify the wetting resistance/vapor flux trade-off due to pore size. We also demonstrate an apparent wetting resistance/vapor flux trade-off due to membrane thickness. These results provide new insight into wetting mechanisms and liquid/vapor flux trade-offs in MD that can guide membrane design for challenging applications.

5b - Laboratory and pilot scale evaluation of membrane distillation for desalination of produced water from unconventional reservoirs

Ritesh Pawar (University of Pittsburgh)* ; Zhewei Zhang (University of Pittsburgh); Radisav Vidic (Pitt)

Increase in hydraulic fracturing of unconventional reservoirs for oil and gas extraction led to an increase in the amount of highly saline produced water generated by this industry. Composition of various chemicals added to fracturing fluid are proprietary and not always known, which makes treating produced water more difficult compared to other conventional wastewaters. Permian basin in Texas is the most productive unconventional play in the U.S. and there is a considerable volume of produced water that is mostly disposed in Class II injection wells. Membrane distillation(MD) uses difference in vapor pressure across membrane as a driving force for treatment of produced water and is not significantly affected by its salinity.

Laboratory-scale membrane distillation experiments with raw produced water showed increase in permeate conductivity with water recovery. Long-term experiment at 50% water recovery revealed that membrane scaling did not affect permeate quality. Several organics compounds and surfactants like 'Nonyl phenol ethoxylate (NPEO)' were detected in both raw water and permeate. Treatment of the synthetic feed solution containing NPEO resulted in permanent wetting of the membrane while the wetting observed when treating produced water can be reversed by simple washing with DI water. Several synthetic solutions were tested in the lab-scale module and it was found that the primary cause for the observed deterioration in permeate quality (i.e., increase in conductivity) was due to the volatilization of ammonia. In addition to these studies, air-gap membrane distillation (AGMD) pilot-scale system equipped with 50% water recovery is currently being operated at a field site using produced water filtered through 25 um bag-house filter and 0.2 um cartridge filter. Comparison of permeate flux and conductivity between the lab-scale and pilot-scale system can provide further validation for the use of MD for treatment of produced water in a full-scale system.

5c - Membrane distillation combined with a refrigerant cycle for enhance performance

Evyatar Shaulsky (Northeastern)*; Zhangxin Wang (Guangdong University of TEchnology); Akshay Deshmukh (MIT); Vicky Karanikola (The University of Arizona); Menachem Elimelech (Yale University)

Thermal desalination technologies consider energy-intensive due to the vast water enthalpy of vaporization but still required for minimal and zero liquid discharge. In this study, we develop a novel desalination technology by using a vapor-compression pump to assist membrane distillation (MD). Comparing the energy efficiencies of the novel heat-pump assisted MD (HPMD), MVC, and conventional MD under similar operating conditions demonstrates that HPMD is an energy-efficient thermal desalination technology. Furthermore, through process modeling, we provide guidelines for HPMD system design and show that the HPMD can theoretically obtain low energy consumption or high gain output ratio, (GOR, of ~60). We conclude by highlighting promising applications of HPMD for MLD/ZLD, enabled by its high energy efficiency, low capital cost, and modularity.

5d - Membrane protein (MP) based nano-porous membranes that transport vapor at high rates while being impermeable to water

Hyeonji Oh (University of Texas at Austin)* ; Yu-Ming Tu (The University of Texas at Austin); Benny Freeman (The University of Texas at Austin); Manish Kumar (University of Texas at Austin)

Encouraged by natural materials, efforts to produce biomimetic membranes by incorporating biological channels into relatively impermeable matrices have been made for years. With the well-defined pore structures of incorporated channels, they surpass the perm-selectivity trade-off which limits the performance of conventional polymer membranes for some applications. However, biomimetic membranes as prepared currently are prone to have defects and require a lot of purified protein channels. In our previous work, we had to perform 6 cycles of membrane protein nanosheet deposition to cover the entire surface of the substrate and achieve the expected selectivity. This time-consuming and high-cost process makes it hard to make biomimetic membranes scalable. Moreover, the use of biomimetic membranes is only limited to the aqueous environment so far because of the potential issues that arise upon dehydration including delamination and damages caused by the removal of surface water from these hydrated materials. In this work, we tried to resolve these challenges by (1) filling the defects in the only 1-cycle deposited OmpF membrane with the UV-cross linkable phospholipid, Diyne PC, and (2) dehydrating the membrane with a low surface tension drying agent, hexamethyldisilazane (HMDS). The resultant membrane demonstrates a water vapor transport rate that is 1.6 and 4 times higher than a commercial breathable fabric, Pertex shield, and the threshold for breathable fabric, respectively. It also shows a very low water permeability of 7 LMH/bar that is 350 times lower than its porous support membrane treated with the same drying agent and 60 times lower than our previous 6-cycle deposited OmpF membrane with a retained solute rejection capacity as demonstrated by a 7-log virus removal capability and an MWCO of 540Da. This shows a great possibility of use as a waterproof breathable protective clothing material.

5e - Surface Patterned Flat-sheet Poly(vinylidene fluoride) Microporous Membrane via Templated Thermally Induced Phase Separation Process: Fabrication and Membrane Distillation Performance

Shouhong Fan (University of Colorado at Boulder)*; Yifu Ding (Membrane Science, Engineering, and Technology (MAST) Center, Department of Mechanical Engineering, University of Colorado Boulder); Kamalesh K Sirkar (New Jersey Institute of Technology); John

Membrane distillation (MD) is promising for recovering fresh water from wastewater of high salinity and temperature. However, the efficiency of a MD process is often limited by both membrane fouling and temperature polarization in the boundary layers. Designed surface patterns are shown beneficial for mitigating the deposition of foulants by inducing higher local shear force on membrane surface during liquid-based separation process. Such a pattern-enhanced mass transfer can potentially improve antifouling performance and reduce temperature polarization in a MD process.

In this presentation, we demonstrate fabrication and MD-process study of a PVDF membrane with designed surface feature. The membranes are fabricated with a lithographically templated thermally induced phase separation (lt-TIPS) process that combines TIPS with soft lithography. Mixed diluent systems with varying ratios of tributyl o-acetylcitrate (ATBC) and diphenyl carbonate (DPC) are used to tune the pore-structure and properties of the patterned PVDF membranes. More specifically, we systematically determined the pattern replication fidelity, pore size, porosity, morphology, mechanical strength, liquid entry pressure (LEP), water contact angle, and pure water flux of the as-fabricated PVDF membranes. We compared the MD process efficiency of the patterned PVDF membrane with non-patterned counterpart using direct contact membrane distillation. The results indicate that co-solvent system is effective at improving the LEP of the PVDF membranes with excellent pattern replication fidelity. The presence of the surface pattern improved the permeation flux during DCMD under identical operation conditions. Using the Wilson plot method, the heat transfer coefficients on feed-side (where the surface pattern interacts with the feed) was measured with varying brine flow rates. The patterned membranes showed improved heat transfer coefficient, compared with non-patterned ones, at lower brine flow rates.

5f - Reduced Fouling Effect on DCMD Desalination by Using Phase-Inversion Synthesized GO-PSF Membrane

Lucy M Camacho (Texas A&M University-Kingsville, Department of Environmental Engineering)* ; Samuel). Olatunji (Texas A&M University-Kingsville)

Membrane distillation (MD) is a promising desalination technology that can reduce energy consumption by introducing solar energy and waste heat. The process of MD can advance by incorporating membranes with engineered hydrophobicity, porosity, and roughness to render high flux and water rejection. The membrane should also have fouling resistance to ensure efficient long-term performance. In this research, the unique properties of graphene oxide (GO) and Polysulfone (PSF) were combined to fabricate robust GO-PSF membranes by phase inversion. The performance and time-dependent fouling resistance of the membranes were tested with semi-pilot direct contact membrane distillation (DCMD) to desalinate reverse osmosis (RO) concentrate. Membrane characterization using contact angle analysis, ATM, and SEM, showed a high porous surface, hydrophobicity, and contact angle, and low roughness. These properties varied with the content of GO, pH of GO sheets, and modification of the phase inversion method. The system provided a permeate flux of 85 L/m2.h, salt rejection of 99.97%, and RO recovery rate of 98%. The membrane fabricated by dry-wet phase inversion, with 1.0% wt. GO, and modified pH of 4.0 of GO sheets showed to be less susceptible to fouling as compared to a commercial polypropylene (PP) membrane. One of the major contributing factors for this behavior was the difference in membrane roughness, which was seven times less for the fabricated GO-PSF membrane as compared to the PP membrane. The anti-fouling properties were observed in both, short- and long-term DCMD experiments. These results are promising to advance the production of MD-dedicated GO-based membranes.

6a - Engineering selective desalination membrane materials via polymer backbone rigidity and functional group position

Geoffrey M Geise (University of Virginia)*

Providing sustainable supplies of purified water and energy is a critical global challenge for the future, and polymer membranes will play a key role in addressing these clear and pressing global needs for water and energy. Polymer membrane-based processes dominate the desalination market because they are more energy efficient than thermal desalination processes, and polymer membranes are crucial components in several rapidly developing power generation and energy storage applications that rely on membranes to control rates of water and/or ion transport. Much remains unknown about the influence of polymer structure on intrinsic water and ion transport properties, and these relationships must be developed to design next generation polymer membrane materials.

For polymers that are of interest for desalination, the water/salt selectivity is a critical property that describes the intrinsic ability of the polymer to transport water as opposed to salt. An observed tradeoff relationship between water/salt permeability selectivity and water permeability suggests that both sorption and diffusion phenomena contribute significantly to water/salt permeability selectivity. This presentation discusses an experimental investigation of the influence of polymer backbone rigidity and functional group position on water/salt selectivity properties of model hydrated polymers. The results suggest that increasing polymer backbone rigidity and distributing hydrophilic functional groups evenly throughout the polymer both drive increases in water/salt selectivity properties that are critical for effective desalination membranes.

6b - Neural Network Model with Evolutionary Algorithm and Bayesian Binary Classification of UF Performance in Pretreatment of Seawater RO Feedwater

Yoram Cohen (Professor)* ; Yang Zhou (UCLA); Muhammad Bilal (University of California, Los Angeles); Han Gu (Orange County Water District)

The performance of ultrafiltration (UF) pretreatment of seawater reverse osmosis (RO) feedwater was explored via an Ensemble Backpropagation Neural Networks (BPNN) model with Alopex Evolutionary Algorithm (AEA). The model was developed for an integrated UF-RO seawater system (18,000 gallons/day capacity) consisting of three multi-bore inside-out hollow-fiber elements and three eight inch spiral-wound seawater RO elements. UF-RO desalination operational data encompassed a wide-range of operating and water quality conditions for both short-term and long-term operational periods, for both fixed and variable coagulant dosing strategies with self-adaptive backwash. Water quality varied considerably over the multi-year study that yielded over 13 million data samples. Based on the dataset, a family of BPNN-AEA models were developed for UF resistance and backwash efficiency using model attributes that were identified and ranked (based on their significance) via multiple feature selection approaches. The family of BPNN-AEA models were assembled into a single ensemble model, following AdaBoost strategy, covering a wide range of operating and water quality conditions, and Bayesian classification to identify UF filtration and backwash operational regimes. Excellent model predictive (and forecasting) performance was achieved for UF fouling resistance and backwash efficiency, even when confronted with temporally variable raw feedwater levels of turbidity (1.5-19 MTU) and chlorophyll a (43-142 mg/L), with R2 in the range of 0.0-0.5 and 0.91-0.97 for post-backwash resistance and backwash efficiency, respectively. The present study suggests that the modeling approach should be useful for (a) UF model-based control, (b) forecast UF performance due to anticipated feedwater quality variations, and (c) quantifying deviations of UF performance from intended baseline performance.

6c - Assessing Performance of Commercial Membranes for Membrane Distillation of Produced Water

Rahul Shevate (University of Houston); Dana N Reed (University of Houston); Devin Shaffer (University of Houston)*

Membrane distillation (MD) is a promising technology for economically treating and reusing high-salinity produced water from oil and gas operations. The primary technological obstacle to the widespread implementation of MD is fouling and wetting of commercial MD membranes. Efforts to develop better-performing MD membranes and to design and operate effective produced water treatment systems are hindered by the current gap in knowledge about the failure mechanisms of commercial MD membranes. This presentation will describe the results of a performance assessment of commercial membrane materials during MD treatment of actual produced water samples from unconventional oil and gas production in the United States. The characteristics of three commercial hydrophobic membrane materials (polyvinylidene difluoride, PVDF; polypropylene, PP; and polytetrafluoroethylene, PTFE) will be presented, as well as the characteristics of the high-salinity produced water used for MD testing. Membrane critical interfacial tension and liquid entry pressure will be introduced as the two measures of produced water-membrane interactions that were strong predictors of membrane performance, and the overall superior performance of PTFE membranes for MD treatment of produced water will be discussed. PTFE membrane fouling rates will be quantified through fouling indices calculated from distillate flux measurements. The composition of observed organic and inorganic membrane foulants will be presented, and the effectiveness of acid and alkaline chemical cleaning for recovering PTFE membrane performance will be reviewed. Membrane characteristics will be correlated to MD performance, and the implications for MD treatment of produced water will be discussed.

6d - Pressure Driven Membrane Filtration for Treating Poultry Processing Wastewater

Ranil Wickramasinghe (University of Arkansas)*

Recovery and reuse of water is essential as the demand for water for industrial and residential uses increases. Here we focus on poultry processing wastewater (PPW). Poultry processing plants use a large amount of water with an average consumption of 26.5 L/bird during primary and secondary processing of live birds to meat. PPW contains proteins, fats, carbohydrates from meat, blood, skin and feathers, resulting in a high biological oxygen demand (BOD) of around 3,000 mg L-1 and chemical oxygen demand (COD) around 6500 mg L-1. Poultry processors are required to treat the wastewater prior to discharge.

Typically, the wastewater from the various unit operations at a poultry processing facility is pooled together and sent to a screening operation to remove larger bone fragments, feather etc. The oil & grease in the wastewater is around 1,500 mg L-1 while the total suspended solids (TSS) is 2,500 mg L-1. The screening step is followed by dissolved air flotation (DAF). The effluent with reduced oil & grease concentrations is generally treated by an activated sludge process (ASP) to reduce the BOD, COD and Total Kjeldahl Nitrogen (TKN). Finally, a second DAF unit is frequently used to separate the biomass from the treated water prior to discharge.

In this work, we investigate the use of a pressure driven membrane filtration to replace the first or second DAF units. Replacing the DAF unit by a membrane filtration process has a number of advantages such as elimination of the use of coagulants and a smaller footprint. Importantly by choosing an appropriate membrane pore size, the membrane could be an absolute barrier to pathogens. A major limitation of membrane technology for treating PPW is membrane fouling. Here we present results for microporous stainless-steel membranes having 0.1 and 0.02 um pore sizes.

6e - City of Lawton's Groundwater Treatment Pilot: Coagulation-Assisted Microfiltration and Side Stream NF/RO

Kamyar Sardari (Garver USA); Michael Watts (Garver USA)*

The City of Lawton currently relies on three (3) surface water reservoirs to meet water supply needs of the City and its wholesale customers. In recent years, the reliability of the City's water supply has come into question due to record drought conditions in southwest Oklahoma. The City enacted an initiative to acquire an additional 5 million gallons per day (MGD) of water supply to bolster the City's water portfolio and maintain desired water supply volumes in the surface water reservoirs during periods of extended drought conditions. A study was performed to evaluate alternative water supply options to supplement the yield from the surface water reservoirs.

Following completion of a groundwater quality testing, it was determined that the most prominent constituents of concern above current regulated Maximum Contaminant Levels (MCLs) in ATH Aquifer included arsenic (As), chloride (Cl), fluoride (F), iron (Fe), and total dissolved solids (TDS). Thus, treatment technologies that can removed these dissolved inorganic constituents were evaluated for the proposed treatment facility including:

1) Nanofiltration (NF) or Reverse Osmosis (RO) (Full Flow)

2) Adsorptive Media Filtration and NF/RO (Side-Stream)

3) Microfiltration (MF) and NF/RO (Side-Stream)

In this work, we aim to present results of the ground water testing for the ATH Aquifer and provide a summary of the treatability study findings conducted to evaluate various treatment techniques mentioned above. In addition, we will present the reasoning for selection of coagulation-assisted NF/RO supported by the treatability study data and provide details of this process train. Furthermore, the authors will provide details of the ongoing pilot trial and touch-on key items included in the pilot protocol.

6f - Membrane Technology vs. Water Prices: Is Water Public or Private Goods?

Albert Kim (U. of Hawai'i)* ; Ahmed Afifi (U. of Hawai'i)

Membrane researchers and engineers investigate the specific physicochemical functionalities of new membrane materials, apply new manufacturing methods in industry, and operate membrane systems/plants on various scales. In the past, the scientific and technological aspects of membranes have primarily received close attention for increasing both rejection and recovery ratios. On the other hand, the commercialization of new membrane materials accompanies the estimation of manufacturing costs, and their real-world applications require a long-term analysis of stable operation and maintenance. For example, in the Middle East, seawater desalination is a national project within strategic plans for sustainability over the coming decades. Further, although people recognize the significance of water scarcity and the unavoidable costs of drinking water production, no specific business model for domestic drinking-water supply is readily available. In economics, goods are classified by their excludable and rivalrous characteristics, often represented as public and private. Everybody needs to pay for a decent quality of drinking water, and no one can privately own large amounts of water. Therefore, it is an unreasonable attempt to include water within one of the predefined categories in standard economics. Instead, water should be classified as a new type of goods, and developing a new paradigm for global water management is an urgent task, along with technological advances of membranes. This work provides in-depth socio-econo-cultural aspects of water to seamlessly link essential entities from the fundamental membrane sciences to the water tariff systems.

7a - Chemically Resistant Thin Film Composite Reverse Osmosis Membranes (V)

Sue J Mecham (NALA Systems)* ; Judy Riffle (NALA Systems)

The dominant structure for reverse osmosis membranes is a layered composite comprised of an ~100-nm thick, non-porous, crosslinked, aromatic polyamide atop a porous polymeric mechanical support with a textile backing. Water flux and the rejection of angstrom size contaminants must be high under the applied pressures to transport large volumes of pure water efficiently. The membranes must not clog easily, and ideally they should last for years with minimum operating downtime devoted to keep them clean. The chemical structure, hydrophilicity and topology of the barrier membrane are critical. NALA focuses on sulfonated polysulfone membranes to avoid biofouling that has plagued polyamide membrane users for decades. These new ionic polysulfones have the unique capability to reject monovalent salts such as NaCl even in the presence of multivalent salts that are in brackish and seawater, whereas previous polymers in this class show poor rejection of salts under such conditions. Their chemical durability, particularly against oxidants such as sodium hypochlorite and hypochlorous acid that are commonly used for disinfection and for cleaning, are outstanding. Accelerated aging tests were conducted in aq. 500 ppm NaOCl/HOCl, then transport measurements were done with 2000 ppm NaCl in the feedwater and with an applied pressure of 400 psi. NALA membranes maintained ~99% NaCl rejection at 50 hours (25,000 ppm-hrs) and >95% rejection after 270 hours (135,000 ppm-hrs). By contrast, commercial TFC controls designed for either brackish water or seawater were not durable to these conditions. This performance suggests that operation using a low level of chlorine would result in a significant reduction in operating costs of >20% by eliminating biofouling to reduce energy costs and dramatically reduce costly CIP operations.

7b - Graphene Oxide Membranes for Industrial Separations

Brandon MacDonald (Via Separations)*

Thermal separations account for approximately 12% of total US energy consumption. Via Separations aims to reduce this, along with resulting greenhouse gas emissions, by developing highly robust membranes and deploying to applications not compatible with existing membrane technologies. This is accomplished through precision chemical modification of graphene oxide in a manner that allows for tailored separation performance while simultaneously imparting excellent chemical and thermal stability.

Via's technology is currently being deployed in pilot systems at pulp and paper facilities where it is incorporated into the black liquor recovery process, the most energy and capital-intensive component of a Kraft pulp mill. Through upconcentration of the high temperature (80°C), high pH (13-14) black liquor solution we are able to dramatically reduce energy costs, while simultaneously breaking production bottlenecks. At a fully commercial scale this technology is projected to eliminate 100 million metric tons of CO_2 emissions by 2050.

7c - Industrial Applications of Membranes in the Membrane Science, Engineering and Technology (MAST) Center at the University of Arkansas

Ranil Wickramasinghe (University of Arkansas)*

The Membrane Science, Engineering and Technology (MAST) Center is a National Science

Foundation Industry and University Cooperative Research Center. The Center's university partners consist of the University of Arkansas, the University of Colorado Boulder, the New Jersey Institute of Technology and the Pennsylvania State University. The MAST Center also collaborates with the R& D Center for Membrane Technology at Chung Yuan University. Being an industry and university co-operative research center the MAST Center's research program is formulated and guided by our numerous sponsors. The Center's research program is divided into a number of research themes.

This presentation focuses on research that is being led by the University of Arkansas Site of the MAST Center which fall into two research themes: biopharmaceutical processing and water purification. Though much of the research is being conducted at the University of Arkansas, the presentation highlights the fact that projects often involve multiple sponsors and faculty at more than one institution. Research in the area of biopharmaceutical processing focuses on bioreactor clarification, virus filtration, membrane adsorbers and protein ultrafiltration. Research in the area of water purification focuses on the development of membrane processes for direct potable reuse of wastewater, treatment of hydraulic fracturing produced water, and poultry production wastewaters.

7d - Development of membrane technologies for the reuse of produced water

Evan Hatakeyama (Chevron)*

In oil & gas extraction operations (upstream), there is an opportunity to reduce local water stress by reusing produced water to support upstream operations, or other beneficial reuse purposes. In general, water treatment process may be needed to remove several constituents including particulate, oil and grease, dissolved organics, dissolved salts, and other constituents.

In one reuse application, water treatment processes convert produced water into desalinated water for steam flood operations. Steam flooding is an enhanced oil recovery (EOR) process that uses steam to heat oil and gas reservoirs to increase production. Pre-treatment, and membrane desalination processes, such as reverse osmosis, may be needed to meet steam flood boiler feed requirements. There are several potential challenges in desalinating produced water for steam flood due to the presence of foulants, high hardness, and high temperatures. In another applications, limited disposal option may require high recovery desalination processes to limit produced water disposal volumes. New process developments are desired to the capital and operational cost of current technologies.

This presentation will cover multiple R&D efforts in membrane technologies to address water treatment challenges in the oil and gas industry. Laboratory and modeling investigation on high temperature reverse osmosis was conducted to understand their application in hot produced water to steam flood. Investigations into membrane distillation (MD) were conducted to understand its potential as a high salinity desalination process. University collaboration to develop membrane technologies to remove organics and trace contaminates was conducted.

7e - Bilayer Aliphatic and Aromatic Polyamide Membranes for High Rejection RO Desalination

John F Thompson (NL Chemical Technology)*; Lars Kohler (NL Chemical Technology); Richard Song (NL Chemical Technology); Jane Li (NL Chemical Technology)

Reverse osmosis (RO) desalination of high salinity brackish water or seawater requires the right balance between high permeability for throughput and high selectivity to produce water for drinking or other uses requiring low solute levels. Such RO membranes are often produced with an ultrathin layer of aromatic polyamide on a microporous polymer that has been cast onto a supporting fabric. For various reasons, any of these layers can contain manufacturing defects that may result in sections of membrane falling out of specification. In particular, the aromatic polyamide layer may contain areas which are too thin or otherwise imperfect, and the microporous support layer may contain pinholes that in turn lead to low solute rejection. In the present approach, we demonstrate a method of coating a flexible top layer of aliphatic polyamide bonded to a base layer of aromatic polyamide for improved solute rejection. This top layer can be used not only to "patch" said defects, but also significantly enhance the rejection characteristics of membranes when manufacturing operations are sound and when no major defects are present. For example, a BWRO membrane could be effectively converted into a reasonably good SWRO membrane using this method. The top layer of aliphatic polyamide consists of an aliphatic polyamine that is water-soluble in its unreacted state. One example of such a polymer is an aminoalkyl-substituted polyacrylate that is rich in primary and/or secondary amine groups. Once bonded to the base layer of aromatic polyamide, the composite bilayer membrane exhibits significantly higher salt rejection characteristics, corresponding to up to a 60 percent reduction in salt passage. A proposed mechanism for this enhanced rejection will be discussed.

7f - Academic partnerships with industry: Creating new membranes from existing products

Jeffrey R McCutcheon (University of Connecticut)*

Industry-academic partnerships can be challenging to forge due to misalignment of incentives and priorities: Industry seeks to increase revenue while academics seek to publish papers and train students. Finding intersections in these incentives can be challenging, but not impossible. One opportunity exists in taking existing products offered by a company and modifying or altering it in such a way that it becomes a different product. This type of work offers the company valuable insight into new product development while giving academics access to a commercial grade material to use as a platform for unique scientific discovery. We present examples of four such interactions:

1. We successfully converted commercially available RO membranes from DuPont into membranes that function for forward osmosis.

2. We successfully altered a 3M microfiltration membrane into a functioning thin film composite membrane for forward osmosis and patented the method.

3. Working with Koch Membrane Systems, we demonstrating the making of a nanofiltration hollow fiber membrane from a braid-reinforced hollow fiber UF membrane.

4. Working with Fraunhofer IKTS and Rauschert North America, we converted ceramic UF membranes into thermally tolerant membrane distillation membranes.

These efforts show one way academia and industry can successfully interact. We will discuss these interactions in the context of what else can be done to facilitate successful industry-academic projects.

8a - Upscaling of Facilitated Transport Membranes for Hydrogen Purification from Coal-Derived Syngas

Yang Han (The Ohio State University)* ; Yutong Yang (The Ohio State University); Ruizhi Pang (The Ohio State University); Winston Ho (OSU)

Amine-containing facilitated transport membranes (FTMs) are of great interest for syngas purification. Various FTMs have shown decent chemical stability at aggressive conditions, but their CO_2/H_2 separation properties are largely limited by the severe carrier saturation at high syngas pressure. Herein, we report a new approach to enhance the CO_2 permeance by manipulating the steric hindrance of the amine carrier. A series of α-aminoacids with different alkyl or hydroxyethyl substituents were deprotonated by 2-(1-piperazinyl)ethylamine, resulting in nonvolatile aminoacid salt carriers with different degrees of steric hindrance. In the presence of moisture, a bulkier alkyl substituent increased the steric hindrance and hence destabilized the carbamate adduct to afford bicarbonate through hydrolysis. Thus, this drastically increased the chemisorption of CO_2. The enhanced CO_2 solubility significantly mitigated the carrier saturation, and an unprecedented CO_2/H_2 selectivity greater than 125 was demonstrated at 107°C and 13.8 bar of CO_2 partial pressure (Type I). As the CO_2 partial pressure reduced to 1.1 bar, a less hindered amine yielded a higher reactive diffusivity of CO_2, resulting in a CO_2 permeance of 217 GPU with a selectivity greater than 268 (Type II). These two types of FTMs were successfully scaled up by a roll-to-roll continuous coating machine. The 14-inch wide scale-up membranes were then rolled into two prototype spiral-wound modules containing 800 and 1600 cm^2 of the Type I and Type II membranes, respectively. The two modules were connected in series, resulting in a hybrid membrane configuration with the Type I membrane treating the syngas near the feed inlet (i.e., high CO_2 partial pressure) and the Type II membrane separating the gas in the proximity of the retentate outlet (i.e., low CO_2 partial pressure). Initial techno-economic analysis indicates that the hybrid membrane configuration can render a H_2 recovery of 99.4% at 90% CO_2 removal

8b - Facilitated Transport Membrane with Ionic Liquid Carrier for CO2 Separation from Air

Yun-Yang Lee (Case Western Reserve University)* ; Drace Penley (Case Western Reserve University); Darrell Jan (NASA); Burcu Gurkan (Case Western Reserve University)

CO2 capture and sequestration technologies are critical for the global goal of carbon neutral as well as carbon negative. While conventional CO2 capture by liquid amine absorbers and solid sorbents are energy intensive, CO2 separation by membrane is a promising technology due to its high energy efficiency and modularity. However, little attention has been given to CO2 separation by membranes from air (direct air capture, DAC) or cabin air (i.e., spacecrafts) where the CO2 concentration, temperature, and humidity level are low, compared to flue gas and other emission sources. This work focuses on a thin film composite type of facilitated transport membrane (FTM) designed specifically for the removal of CO2 from air. The selective layer is an ionic liquid (IL)-poly(IL) gel nanoconfined within the deposited graphene oxide nanosheet framework (GONF). High performance of CO2 carrier-facilitated transport at conditions relevant to DAC (3090 GPU and CO2/N2 selectivity of 1180) and cabin air (620 GPU and CO2/N2 selectivity of 250) was demonstrated at 295 K and 40% RH. The CO2/O2 selectivities for air (410 ppm CO2) and cabin air (2500 ppm CO2) feed were measured to be about 350 and 40, respectively, at 313 K.

The fabricated membranes were characterized by IR, XRD, and SEM-EDS, and verified the PIL/IL-impregnated GONF layer over a PES/PET support. The reaction mechanism by which CO2 binds to the carrier and the associated reaction enthalpies are examined by NMR and density functional theory. The diffusivity of the CO2 complexes are studied by DOSY NMR. The transport of CO2-complex was revealed to be the bottleneck of the overall process. The fabricated membranes were subject to continuous separation for two weeks (410 ppm CO2 at 295 K) without performance decay, thus demonstrating stability. This study represents one of the few FTMs for DAC with a superior permeation-selectivity tradeoff.

8c - Molecular Design of high-performance ionenes and ionic composites for gas separation membranes

Kathryn E O'Harra (University of Alabama)* ; Jason Bara (University of Alabama)

We have developed a library of high-performance (HP) ionenes, polymers containing ionic groups within the backbone (as opposed to pendants). The design of these materials merges structural elements typically associated with state-of-the-art gas-separation membranes with high-performance engineering polymers such as polyimides, polyamides, and PEEK. The incorporated functional groups and high free volume moieties (such as Tröger's Base) are regularly spaced by imidazolium cations paired with fluorinated, delocalized anions (i.e. Tf2N). These HP-ionenes are thoroughly characterized in order to understand the underlying relationships between HP-ionene structure and key properties such as thermal transitions and gas permeability. Furthermore, the thermal and mechanical properties as well as gas transport behaviors can be tuned, based on the integrated structural and functional features and by the addition of free ionic liquid (IL) into the HP-ionene matrix. These neat and composite membranes show great promise for separation of CO_2 from gas mixtures.

8d - CO2 capture from gasification streams: Gen-2 ProteusTMmembrane development and module field test results

Witopo Salim (Membrane Technology & Research Inc.)* ; Karl Amo (Membrane Technology & Research Inc.); Richard Baker (Membrane Technology & Research Inc.); Vincent Batoon (Membrane Technology & Research Inc.); Carlos Casillas (Membrane Technology & Resear

Gasification of coal, biomass or other feedstocks is widely used in industry for the production of hydrogen, chemicals, and power. In a carbon-constrained world, gas separation is an increasing important step to maintain the economic viability of these processes with CO_2 capture. Membrane Technology and Research, Inc. previously proposed a dual membrane process for CO_2 capture from gasification processes, including ProteusTM membranes for H_2 removal or purification and PolarisTM membranes for CO_2 recovery. This dual-membrane process can achieve a H_2 recovery of 99% with >90% CO_2 capture while producing high purity liquid CO_2 at an energy cost lower than conventional technology.

This presentation focuses on the development and field test of the H2-selective Gen-2 Proteus membranes and modules. Field test results of the Gen-1 Proteus bench-scale modules with a syngas slipstream at the National Carbon Capture Center (NCCC) in Wilsonville, AL showed promising results over 9,000 hours of testing. Membranes and modules showed H_2/CO_2 selectivities of 15 or more when treating real syngas containing up to 800 ppm H_2S. Based on these results, a preliminary techno-economic analysis of the MTR dual-membrane process resulted in a CO_2 capture cost of $49/tonne of CO_2, which is approximately 20% lower than the cost for the conventional absorption technology, SelexolTM. The Gen-2 Proteus membrane under development at MTR will further reduce costs with an improved upper temperature limit and selectivity compared to Gen-1 Proteus. The Gen-2 Proteus membrane and module development will be discussed and module field test results with real syngas at Energy and Environmental Research Center (EERC) of University of North Dakota will be presented. Pressure and temperature parametric testing of the Gen-2 module with real syngas containing up to 1400 ppm H_2S was completed with field test results in excellent agreement with mixed gas test results conducted at MTR.

8e - Synergistic Thermolabile Cross-linking and Low-temperature Carbonization to Create Sub-3.3 Ultramicropores for Membrane H_2/CO_2 Separation

LEIQING HU (University at Buffalo)* ; Haiqing Lin (The State University of New York at Buffalo)

Membrane technology with superior H_2/CO_2 separation properties at high temperatures is of great importance for pre-combustion CO_2 capture. Carbon molecular sieves (CMS) membranes prepared by carbonization of polymers contain ultramicropores with a strong molecular-sieving ability and are attractive for high-temperature gas separations. However, polymers need to be carbonized at extremely high temperatures (900 – 1200°C) to achieve sub-3.3 ultramicroporous channels required for H_2/CO_2 separation, which makes them brittle and impractical for industrial applications. Herein, we demonstrate that polymers can be first doped with thermolabile cross-linkers before low-temperature carbonization to retain the polymer processability and achieve superior H_2/CO_2 separation properties. Specifically, polybenzimidazole (PBI) is cross-linked with pyrophosphoric acid (PPA) via H-bonding and proton transfer before carbonization at <600°C. PBI is a leading material for H_2/CO_2 separation with H_2 permeability of 27 and H_2/CO_2 selectivity of 16 at 150 oC because of the strong size-sieving ability. The PPA doping and pyrolysis increase the H_2 permeability to 140 Barrer and H_2/CO_2 selectivity to 58, moving the properties above Robeson's upper bound at 150°C. Moreover, when challenged with simulated syngas contaiing water vapor, the CMS membranes demonstrate stable mixed-gas selectivity for 105 h. We will thoroughly describe the physical characterization of the CMS membranes, aiming to derive the structure/property relationship. In summary, the synergistic doping and carbonization render a facile and effective way to manipulate micropore size and porosity in CMS membranes for desirable molecular sieving ability.

8f - Nano-Hybrid Thin Film Composite Carbon Molecular Sieve Membranes based on a Polymer of Intrinsic Microporosity

Wojciech Ogieglo (KAUST)* ; Tiara Puspasari (KAUST); Khalid Hazazi (KAUST); Mrinal Hota (KAUST); Husam N Alshareef (KAUST); Ingo Pinnau (KAUST)

Carbon molecular sieve (CMS) membranes promise to deliver outstanding molecular separation properties combined with a high chemical and plasticization resistance unmatched by currently deployed polymeric materials. The main challenge, however, remains in transforming them into practical thin film composites with high enough permeances while simultaneously avoiding the development of defects and preventing substantial mass transport resistance increase due to physical aging[1, 2].

Recently, we have introduced nano-hybrid carbon molecular sieve membranes fabricated by pyrolyzing a polymer of intrinsic microporosity (PIM) precursor modified by vapor phase infiltration (VPI)[3]. In the process of VPI a metalorganic precursor (trimethylaluminium, TMA) first diffuses into the high free volume matrix of the PIM to form a complex with its functional (carbonyl) groups. Afterwards, water vapor selectively and locally oxidizes the TMA forming nano-disperesed Al_2O_3 within the PIM matrix. Subsequent inert-atmosphere pyrolysis leads to the formation of Al_2O_3-doped thin film composite CMS membranes with excellent molecular separation properties positioned in the vicinity or above the polymeric upper bounds for a number of technologically important gas pairs, e.g. CO_2/CH_4, O_2/N_2, and H_2/N_2. In addition, we show that the nano-hybrid CMS membranes possess a distinct physical aging signature that may lead to a significant improvement of the long-term performance.

We will also present some results related to attempting to scale-up the nano-hybrid CMS membranes with the utilization of stainless steel porous fiber supports.

References:

[1] Ogieglo, W.; Puspasari, T.; Ma, X.; Pinnau, I. Journal of Membrane Science 2020, 597.

[2] Hazazi, K.; Ma, X.; Wang, Y.; Ogieglo, W.; Alhazmi, A.; Han, Y.; Pinnau, I., Journal of Membrane Science 2019, 585, 1-9.

[3] Ogieglo, W.; Puspasari, T.; Hota, M. K.; Wehbe, N.; Alshareef, H. N.; Pinnau, I., Materials Today Nano 2020, 9.

9a - RO Membrane Compaction and Permeate Carrier Embossing at High and Ultra-High Pressure

Jishan Wu (UCLA)* ; Eric Hoek (UCLA)

Over the past half century, reverse osmosis (RO) has grown from a laboratory scale concept to be the most cost effective and versatile water treatment technology available. The combination of water scarcity, impaired fresh surface and ground water quality, and sustainability imperatives is driving RO system designs towards minimal and zero liquid discharge (M/ZLD). If novel RO membranes can be developed to tolerate ultra-high pressures (up to 200 bar = 2,900 psi), MLD/ZLD can be realized at a significantly lower energy and cost than conventional thermal distillation processes. However, the membrane compactions and permeate carrier embossing on the membrane fabric significantly increase at such high pressures, which eventually damage the membrane performance, minimize the productivity, and increase the overall cost. Herein, brackish water RO (BWRO), seawater RO (SWRO) and high-pressure RO (HPRO) membranes performance (e.g., flux, rejection, and compaction) are discriminated within each composite layer (i.e., coating film, support membrane, fabric) at pressures from 200 psi to 3000 psi. Damage on the membrane surface to different degrees with different permeate carriers are observed at high and ultra-high pressures. The permeability of all tested membranes declines with increasing pressure, whereas rejection behaves differently for different membrane types. FIB-SEM images suggest collapse of support membrane skin layer pores plays a major role in performance decline due to physical compaction. Moreover, the effects of membrane "embossing" by different permeate carriers, including BWRO, SWRO, HPRO flat-sheet membranes, traditional tricot mesh permeate carriers, and novel "cloth-like" permeate carriers are reported.

9b - Catalytic Membranes for Integrated CO2 Capture and Conversion

Casey O'Brien (University of Notre Dame)* ; Renxi Jin (University of Notre Dame); Hui Xu (University of Notre Dame); Sarah Pate (University of Notre Dame); Justin Easa (University of Notre Dame)

State-of-the-art CO2 capture technologies are sorbent-based technologies that are extremely costly because they require inefficient process steps to release CO2 and regenerate the sorbent, and to compress and transport CO2 for storage or utilization. We are developing an integrated CO2 capture and conversion concept that addresses these challenges and potentially enables commercially viable CO2 capture by integrating CO2 capture and CO2 conversion to value-added chemicals in a single unit process operated continuously at mild conditions.

Specifically, the innovative enabling technology is an amine-functionalized polymeric membrane that captures CO2 from dilute sources and catalytically converts it with epoxides to produce cyclic carbonates. The multi-functional membrane acts as both the CO2 capture and conversion medium, thereby eliminating the energy intensive step of regeneration and CO2 pipeline infrastructure, and enabling continuous operation at mild temperatures.

In this talk, I will demonstrate the proof-of-concept of the integrated CO2 capture and conversion concept by showing that amine-functionalized polymeric membranes installed in a benchtop membrane reactor system can capture CO2 from dilute sources and convert it with epoxides to form cyclic carbonates in a separate stream. I will also describe how the CO2 separation performance and catalytic activity of the amine-functionalized membranes can be tuned. Finally, I will describe unique operando spectroscopy tools that we have developed to probe the fundamental mechanisms of the highly integrated catalytic membrane separation process.

9c - What do fuel cells, gas separation, and reverse osmosis have in common? Unifying the conversation across membrane applications to enable cross-pollination

Sarah M Dischinger (Lawrence Berkeley National Laboratory)* ; Ryan S Kingsbury (University of North Carolina at Chapel Hill); David Vermaas (TU Delft); Daniel J Miller (Lawrence Berkeley National Laboratory)

Dense polymeric membranes enable a diverse array of chemical separations for power generation, water purification, and chemical manufacturing. Despite widespread interest, terminology and performance metrics vary considerably, limiting the cross-pollination of advances between fields. In this presentation, we will describe a universal framework for comparing the performance of membrane-based separations that is grounded in a first-principles approach to transport, and subsequently utilize this framework to assess the challenges facing multiple membrane applications.

Solute transport across a membrane is the net result of a driving force (i.e., the electrochemical potential gradient) and a resistance due to the membrane properties; the physicochemical differences of the solutes themselves give rise to different transport rates and thereby result in separation. Therefore, these physicochemical differences determine the extent to which driving force and membrane properties can be leveraged to accomplish the separation. In contextualizing the universal framework within the physicochemical properties of the solutes, we will put forth that the design of membrane separations for new applications must be centered around the physicochemical differences of the solutes involved.

The universal framework enables us to explore trends in membrane separation performance from literature data spanning 28 separations across nine applications. We will discuss the importance of solute rejection in achieving high membrane selectivity and the opportunities presented by electric field-driven separations. Recognizing that poor rejection of small uncharged solutes is a challenge confronting fuel cells, gas separations, and reverse osmosis, we will discuss opportunities for cross-pollination between these applications. Based on our analysis, we will offer a perspective on the barriers to and opportunities for development of separations utilizing dense polymeric membranes.

9d - Polyol-Functionalized Polyether Membranes for Selective Removal of Boric Acid

Matthew R Landsman (University of Texas at Austin)*; Frederick Rivers (University of Texas at Austin); Benjamin Pedretti (University of Texas at Austin); Gregory Su (Lawrence Berkeley National Laboratory); Benny Freeman (The University of Texas at Austin)

Conventional desalination membranes provide poor removal of boric acid, which often necessitates multi-stage treatment trains to achieve desired boron removal for end uses such as irrigation. The development of membranes that effectively remove boron during single-pass permeation could expand our portfolio of energy-efficient water reuse technologies. One approach to designing membranes with high removal of boric acid is via incorporation of chelating ligands that selectively sorb boron. In this study, a versatile membrane platform based on poly(glycidyl glycidyl ether) (PGGE) was functionalized with N-methyl-D-glucamine (NMDG), a polyol known to interact selectively with boron. Boron sorption isotherms in PGGE and PGGE-NMDG were well-described by dual mode isotherms, presumably due to absorption of boron in the polymer matrix and additional specific interactions between boron and functional groups in the membranes. Experiments performed using buffered solutions of boric acid (pH 7.0) indicated that borate-NMDG complexes form in PGGE-NMDG membranes over all tested concentrations. Boron sorption in PGGE and PGGE-NMDG membranes was reversible over four cycles of acid regeneration without a significant loss of boron uptake capacity. Diffusion of boron in PGGE and PGGE-NMDG membranes exhibited two-stage sorption behavior, which may result from specific interactions such as chelation by NMDG groups and hydrogen bonding with epoxide, ether, and amine groups. Ongoing work is investigating diffusion mechanisms to describe two-stage sorption behavior and employing spectroscopic techniques to elucidate specific complexes between boron and NMDG. This research highlights several important properties of ligand-functionalized membranes that can be leveraged to enhance removal of challenging solutes such as boric acid during water purification.

9e - Optimization of a concentration gradient battery using an osmotic ballast to enhance saltwater-based energy storage

Holly M Haflich (University of North Carolina- Chapel Hill)* ; Mikayla D Armstrong (University of North Carolina at Chapel Hill); Fei Liu (University of North Carolina at Chapel Hill); Orlando Coronell (University of North Carolina at Chapel Hill)

Concentration gradient batteries (CGB) apply principles of electrodialysis (ED) and reverse electrodialysis (RED) for energy storage. A CGB is charged and discharged using ED and RED, respectively. The salinity gradient and resulting osmotic pressure differences between concentrate and dilute chambers in the CGB lead to unwanted osmotic water transport which hinders the CGB energy efficiency and power density. Previous work showed that adding an osmotic ballast to the dilute chamber limited osmosis and improved efficiency. Past work has modeled effects of CGB parameters on performance, but there is limited experimental validation and no investigation of CGB parameters with a ballast. Thus, knowledge gaps exist regarding how the physical and operating parameters of the CGB influence performance and how the ballast alters effects of CGB parameters. Accordingly, the goal of this study was to evaluate the effects of selected CGB physical and operational parameters on performance toward realizing an optimum CGB design. We evaluated effects of ion exchange membrane (IEM) properties, current density, spacer thickness, salinity gradient, ballast loading, and flowrate on power density and energy-efficiency. We used our results to evaluate the performance of an optimized CGB. Results show that effects of IEM properties, salinity gradient, and current density were consistent with or without the ballast, whereas effects of spacer thickness and flowrate were not. Accounting for hydrodynamic power loss, power density ranged from -236 to 91.5 mW/m2. The ballast improved round-trip energy-efficiency from 12.7 to 34.6% and allowed the CGB to maintain performance during multiple charge-discharge cycles. To further optimize the CGB, the electrode rinse solution, charge/discharge time, and electrolyte solutions should be studied. This work improves understanding of how adjusting CGB parameters informs performance and may inform methods to hinder osmosis in other IEM-based processes.

9f - Multi-Objective Optimization of the Economic Feasibility for Mobile On-Site Oil and Gas Produced Water Desalination and Reuse

Garrett M Cole (Colorado State University)* ; Cristian Robbins (Colorado State University); Tiezheng Tong (Colorado State University); Todd Bandhauer (Colorado State University); Jason Quinn (Colorado State University)

Development of unconventional oil and gas has resulted in large volumes of flowback and produced (FP) water that require careful handling to minimize environmental and human health risks due to high concentrations of salts and other contaminants. Common practice is to truck the wastewater from producing well sites to Environmental Protection Agency class II underground injection control (UIC) wells. As an alternative, Reverse Osmosis (RO) and Membrane Distillation (MD) deployed for on-site desalination followed by surface water discharge of the water product reduce the volume of wastewater requiring transport to UIC wells and transportation cost while additionally alleviating strain on water supplies in arid regions. In contrast to centralized desalination, on-site desalination places importance on scalable treatment capacities and the mobility of desalination equipment. Scalable treatment capacity is important because of the extreme decay in FP water volumes over the lifetime of a producing well. One way to scale treatment capacity is to run multiple packaged RO or MD units of a standardized capacity in series, moving them to new producing wells as demand fluctuates. This work explores different RO and MD FP water management strategies in Weld County, Colorado, to determine a set of Pareto optimal produced water management strategies from a techno-economic and environmental perspective optimizing economics and water reclamation. Results show that properly scaling treatment capacity can reduce management cost by as much as 2.46 $/m3 compared to treatment at a fixed capacity during the one year study, and helps make RO and MD competitive with injection. Generally, optimal management strategies for Weld County, utilized packaged RO and MD units of 100 m3/d capacity with deployment location reevaluated on a 1-6 month basis.

10a - Unifying the Pore-flow and Solution-diffusion descriptions for solvent transport through swollen, non-porous membranes

Varun H Hegde (University of California Santa Barbara)* ; Todd Squires (University of California Santa Barbara); Michael Doherty (University of California Santa Barbara)

Solvent transport through membranes is typically treated using one of two models. The pore flow model, which describes fluid flows through rigid membranes with fixed pores, treats permeation as a pressure-driven flow governed by the equations of fluid mechanics. The solution-diffusion model, by contrast, describes solvent flux through homogeneous, non-porous membranes. The solution-diffusion approach treats solvent transport as binary diffusion, where transmembrane differences in solvent activity establish concentration gradients within the membrane, driving a gradient-driven diffusive flux.

A long-standing and ongoing controversy surrounds which model is appropriate to describe solvent transport through highly swollen membranes. Because the liquid forms a continuous phase that spans the membrane, pressure gradients within the liquid should drive a solvent flow, as described by pore flow. At the same time, highly-swollen gels are osmotically active, whose configuration is determined by diffusive fluctuations of the polymer relative to the solvent, as described by solution-diffusion.

Despite this apparent discrepancy, we argue that both approaches are valid, representing complementary approaches to the same phenomenon, and make quantitatively identical predictions. We do so by adapting the fluid-solid model, which was developed to describe the dynamics of gels and soil consolidation, and which treats the solvent and matrix as separate phases. Like the pore-flow model, the fluid-solid model uses a momentum balance to describe solvent flow, yet additionally tracks membrane deformations due to solvent-matrix drag. The fluctuation-dissipation theorem connects the (deterministic) fluid-solid model directly to the (stochastic) solution-diffusion model, yet remains consistent with the Pore flow model's mechanical approach. Apparent discrepancies between the predictions of the two classical models are shown to be resolved.

10b - Molecular Mechanisms of Ion Selectivity in Nanoporous Polymeric Membranes

Cody Ritt (Yale University)* ; Mingjie Liu (Massachusetts Institute of Technology); Tuan Anh Pham (Lawrence Livermore National Laboratory); Razi Epsztein (Technion - Israel Institute of Technology); Heather Kulik (Massachusetts Institute of Technology); Me

Designing single-species selective membranes for precision separations applications requires a fundamental understanding of the molecular interactions governing solute transport. In this study, we comprehensively assess molecular-level features that influence the separation of 18 sodium-based anions by nanoporous cellulose acetate membranes. Our analysis identifies the limited role of using bulk solvation characteristics (e.g., hydration energy and hydrated radius) for explaining transmembrane ion transport while highlighting the critical need to consider entropy to holistically study transport thermodynamics. Specifically, we find that entropy accounts for more than half of the total free energy barrier (ΔG‡) for most of the anions studied here. Entropy–enthalpy compensation (EEC) is observed in the transport thermodynamics spanning a range of 40 kJ/mol, leading to a ΔG‡ variation of only ~8 kJ/mol across all anions. With advanced machine learning techniques, we determined important descriptors for the anion energetic barriers from a set of 126 features sourced from cheminformatics, literature, and first-principles simulations conducted herein. Notably, electrostatic features account for 75% of the normalized absolute coefficients (∑β) used to describe ΔG‡, despite the relatively uncharged state of cellulose acetate. While ion-membrane interactions contributed significantly to the entropic barriers (∑β = 45%), their reduced role in ΔG‡ suggest that EEC mitigates the importance of ion binding with the polymer network to the overall transport. The learned ΔG‡ model is able to accurately predict the separation of an unlearned monovalent anion, azide (N3-, error < 1%). Collectively, our findings call for a paradigm shift in our approach to studying ion transport across nanoporous membranes and suggest that research efforts should focus on designing confined environments, rather than specific materials or functionalities, best-suited for target ions.

10c - A Theoretical Methodology of Mobile Carrier Evaluation in Facilitated Transport Membranes

Xuepeng Deng (The Ohio State University)* ; Changlong Zou (The Ohio State University); Yang Han (The Ohio State University); Li-Chiang Lin (The Ohio State University); Winston Ho (OSU)

Incorporating molecular amines as mobile carriers (MCs) in facilitated transport membranes (FTMs) significantly enhances the CO_2 permeance and CO_2/N_2 selectivity. Searching for better-performing MCs from the vast chemical space of amines can be demanding using solely experimental techniques. To facilitate the discovery of promising MC candidates, we have developed a computational methodology by employing density functional theory (DFT) calculations, molecular dynamics (MD) simulations, and Monte Carlo simulations to systematically evaluate MCs from the perspectives of the amine–CO_2 reaction chemistry, diffusivities of carriers and gases, and N_2 sorption. This presentation focuses on a case study where the proposed methodology was applied to analyze two well-known MCs, piperazine glycinate (PZ-Gly) and 2-(1-piperazinyl)ethylamine sarcosinate (PZEA-Sar) along with the fixed-site carrier. DFT calculations revealed that PZEA-Sar exhibited a higher CO_2 loading capacity as compared to PZ-Gly in line with the higher CO_2 permeance of the former. The water uptake of the FTM was also shown to influence the FTM performance significantly. MD simulations showed that the diffusivities of both CO_2 and MCs increased drastically with the water uptake. The N_2 solubility calculated using Monte Carlo methods also varied with the water uptake. Experimental measurements revealed that the addition of PZEA-Sar substantially increased the water uptake, which consequently led to faster diffusion of both CO_2 and MCs. This observation agrees well with the enhanced CO_2 permeance observed in FTMs with MCs. Our methodology also accurately described the N_2 permeance of FTMs with different carrier compositions. The proposed methodology has been extended to study other promising MC candidates. To the best of our knowledge, our approach is the first of its kind to evaluate MCs in FTMs at the molecular level and is easily transferable to other membrane-based technologies.

10d - Three-dimensional Flows and Dean Vortices in Membrane Distillation Systems

Ankun Wang (Stanford University)* ; Ilenia Battiato (Stanford University); Meagan Mauter (Stanford Univerisity); Bowen Ling (Stanford University); Alexander V Dudchenko (SLAC National Accelerator Laboratory)

Membrane distillation (MD) technology is an emerging and promising technology for water desalination. High fidelity numerical simulations are important to study fluid flow, mass, and heat transfer in MD processes and to guide the design of, and improve, MD water recovery and energy efficiency. Due to their computational burden, generally computational fluid dynamics (CFD) simulations in the context of MD are conducted in 2D geometries and employed to parametrize Sherwood and Nusselt correlations, later used in the design at the module-scale. However, the simplifying hypothesis of 2D flow and transport can dramatically impact the accuracy of the predictions of mass and heat transfer, as well as momentum transport, in complex geometries. Specifically, when boundary effects are predominant, fully 3D simulations may be necessary to accurately estimate Sherwood and Nusselt correlations in realistic setups. In this work, we numerically study the impact of boundary (entrance/exit) effects on mass and heat transfer in MD systems. Specifically, we study the impact that Dean vortices - inherently three-dimensional flow structures absent in two-dimensions – have on the accuracy of traditional Nusselt and Sherwood correlations in MD modules with non-negligible entrance/3D effects. Finally, we investigate some of the topological features of Dean vortices, e.g. their size and number, in terms of (i) relevant scale-ratios of the MD module, (ii) Reynolds number, and (iii) the dynamic correlation between heat/mass transport parameters and temperature and salt concentrations.

10e - Predicting the transport of soft droplets in porous media from measurable emergent properties

Guillaume G Lostec (CU Boulder)* ; John Pellegrino (University of Colorado); Franck Vernerey (CU Boulder)

Understanding the transport of soft particles (droplets, colloids, vesicles, cells) in porous media is relevant to drug delivery, medical diagnostics, water treatment as well as food and cosmetic industries. Toward this objective we previously proposed a multiscale modeling approach, connecting the mechanics of particles deforming through individual pores to the topology of the porous medium. The experimental validation of this model and its predictive potential are yet limited by the lack of a calibration procedure. Indeed, the model micro structure is highly idealized and the parameters describing its geometry cannot be directly measured on a real membrane. To address this issue, we present a method to find the value of the model's parameters such that the idealized geometry can be considered equivalent, in terms of soft particles transport, to a real porous medium. We first review the characteristics of the model with an emphasis on the input parameters and their influence on particles transport. Then, we introduce our calibration process based on matching the value of a set of measurable emergent properties in the model with a real membrane. We present a phase diagram predicting the transport of soft particles through a specific membrane and how it can be obtained with the model calibration. Finally, we show that the model predicts three different permeation regimes depending on the value of the emergent properties under the form of a second phase diagram.

10f - Origins of cation-cation selectivity in crown ether-functionalized polymer membranes

Everett S Zofchak (The University of Texas at Austin)* ; Zidan Zhang (The University of Texas at Austin); Bill Wheatle (The University of Texas at Austin); Rahul Sujanani (The University of Texas at Austin); Sam Warnock (University of California, Santa Bar

Selective separation of monovalent cationic species from complex mixtures is an industrially relevant procedure necessary for the recovery of many commodity materials, such as lithium from salt brines. Unfortunately, most conventional membranes lack selectivity between monovalent ions, rendering their use in such applications infeasible. One approach to overcoming selectivity limitations is to incorporate ligands into polymer matrices which specifically interact with target cations in an aqueous environment. In this work, we assess how incorporating crown ethers, which form host-guest complexes with monovalent cations, into polynorbornene networks impacts the selective partitioning and diffusion of alkali cations. For the case of a 12-Crown-4-functionalized membrane, atomistic molecular dynamics simulations reveal that an unprecedented LiCl/NaCl permeability selectivity of ~ 2.3 arises from strong, favorable interactions between 12-Crown-4 and Na+ ions. These interactions, despite favoring NaCl partitioning into the membrane, sufficiently reduce NaCl diffusivity to permit selective permeation of LiCl. Further simulation studies reveal that a maximum in diffusivity selectivity occurs at an intermediate water volume fraction due to the competition between ion-ligand complexation and free volume size/distribution. Finally, we reveal that ion diffusivities, when scaled by their respective solution diffusivities and free ion fractions, collapse onto an almost universal curve depending on solvent volume fraction. These simulation results provide critical molecular-level insight into the interplay between membrane chemistry and monovalent ion selectivity, aiding in the rational design of selective membranes for resource recovery.

11a - Forward osmosis membrane use for dialysate regeneration to enable portable kidney dialysis

Guozheng Shao (Univ. of Wash.); Bruce Hinds (Univ. of Washington)*

Kidney disease effects 37 million people in the US with end stage renal failure requiring life-saving kidney dialysis that consumes 7% of the US annual medicade budget. In-center hemodialysis is the most prevalent (90%) but suffers from poor 5 year morbidity of 65% and severely reduced quality of life due to being tethered to a fixed dialyzer for 4 hours thrice weekly with associated non-physiological toxin build up. Ideal is to develop a small portable device that operates continuously, however this is prevented due to conventional dialysis session requiring 120 kg of purified water with added electrolytes. This can be reduced to about 1kg if the dialysate can be regenerated. We had previously demonstrated that a UV-TiO_2 based Photo-Oxidation Urea Removal (POUR) system [DOI: 10.1021/acsanm.9b00709] removed 14gr/day of urea to regenerate dialysate. However, careful analysis of spent dialysate species showed that the POUR unit would have to be 3 times larger to oxidize other organics such as glucose and amino acids. Hence a urea selective membrane would be required to increase the POUR unit efficiency. Lack of urea rejection with osmotic membranes (RO/FO) has been a long-standing problem in wastewater treatment, but this urea selectivity directly benefits dialysate regeneration. Using physiological mixtures of prevalent dialysate species, a compact RO unit showed rejections of 24% for urea, 79% for creatinine, 98% alanine, and greater than 99% for 6 other highly prevalent uremic species. Operation in FO mode with urea diffusion into POUR unit loop is superior since it uses low pressure pumps. Commercially available hollow fiber FO membranes (Aquaporin, 2.3m^2) with a 10mM concentration gradient had a urea flux of 30gr/day that was more than twice that required and corresponded to a urea mass transfer coefficient of 4x10^-7 m/s. The FO-POUR system was validated using human spent dialysate and met required 14gr/day urea removal in a recirculating dialysate.

11b - Advancing Microporous Membranes for Mask and Filter Applications Towards Aerosol Capture and Coronavirus Deactivation

Dibakar Bhattacharyya (U. Kentucky); Rollie G Mills (University of Kentucky)* ; Ronald Vogler (University of Kentucky); Matthew Bernard (University of Kentucky); Jacob Concolino (University of Kentucky); Yinai Wei (University of Kentucky); Todd Hastings

Aerosolized particles are the primary means of various virus transmission, thus furthering the technological development of respiratory face masks and enclosed environment filters is important for virus spread. In this work, the adaptation of various water-filtration membranes to air filtration was investigated, as well as including membrane functionalization with enzymes that can deactivate spike glycoproteins (SGPs) in low humidity environments. The air flux of various water filtration membranes (PVDF400-MF, PS35-UF, polycarbonate-MF, etc.) were evaluated at human breathing pressures (0.07-0.08 bar) and compared to commercial N95 and surgical masks. A model PVDF membrane design with appropriate thickness, porosity and pore size was proposed to attain high air permeability needs. Aerosolized polystyrene latex (PSL) particles of similar size to the SARS-CoV-2 virus (100-120 nm) were utilized in this study to test filtration efficiency and particle capture of membranes and commercial masks. For the immobilization of Subtilisin Carlsberg, a stable protease enzyme commonly used in laundry detergent, PVDF400 membranes were functionalized with poly (methacrylic acid) (PMAA) to enhance the immobilization of enzymes in a batch or convective mode. Subtilisin Carlsberg was proven to deactivate SGP of SARS-CoV-2 in solution phase with the use of Differential Scanning Calorimetry (DSC). Using Sypro Orange, a hydrophobic-binding fluorescent dye, Subtilisin-functionalized PMAA-PVDF membranes was proven to denature SGP quickly with minimal hydration (1.35 µl water/cm2 of membrane surface). Taken together, these results demonstrate that water-filtration membranes can be applied to air-filtration applications and, when functionalized with enzymes, can deactivate SGPs in low-hydration environments, thus creating exciting technology for the development of next-generation respiratory face masks and filters. This research is supported by the NSF-RAPID program and by NIEHS.

11c - 3D Printed Adsorbers for Capturing Chemotherapy Drugs before They Spread Through the Body

Hee Jeung Oh (Pennsylvania State University)*

Due to longer life expectancies, the prevalence of age-related diseases is increasing rapidly, and the need for developing biomedical devices that can solve big health problems is similarly greater. Inspired by absorption columns, which are routinely used in industry to remove pollutants from chemical streams, this research describes the design of biomedical devices for capturing unwanted toxins in the body. One significant benefit of using polymer membranes is their tunable binding affinity to target molecules using specific chemical, physical, or biological features. One example is using properly designed polymers to remove cancer chemotherapy drugs that are not taken up by the target tumor during chemotherapy to reduce the drugs' toxic side effects.

Cancer is becoming the leading cause of death in most developed nations. Despite efforts to develop targeted and personalized cancer therapeutics, dosing of the cancer chemotherapeutics is limited by toxic side effects. During intra-arterial chemotherapy infusion to a target organ, typically, more than 50-90% of the injected drug is not trapped in the target organ and bypasses the tumor to general circulation, causing toxicities in distant locations.

In the context of reducing the toxicity of chemotherapy, we have designed, built, and deployed porous adsorbers for capturing chemotherapy drugs before they spread through the body. The porosity was obtained by 3D printing of lattice structures. The surface of porous cylinders was coated with an ion-containing nanostructured block polymer which is responsible for capturing doxorubicin, a widely used chemotherapy drug with significant toxic side effects. Using a swine model, our initial design enables the capture of 69 % of the administered drug without any adverse effects. Additional improvement may be obtained by changing the chemical composition of the selective membrane layer and controlling the lattice structure and size with elastomers.

11d - Towards an Artificial Kidney

Jamie Hestekin (University of Arkansas)* ; Christa Hestekin (University of Arkansas); Ira Kurtz (UCLA (Nephrology)); Roland Ludlow (US Kidney Research Corp)

In the past 40 years there has been little in the way of advances for end stage renal disease. Although dialysis keeps patients alive, it tethers them to a location by requiring dialysis up to 3 times a week. There are several technologies going on to try to increase the freedom of dialysis patients by making a portable dialysis device. However, even these devices require dialysate that must be adjusted for patient needs. In a partnership with UCLA and the US Kidney Research Corporation, the University of Arkansas has developed an artificial kidney device that produces urine without the use of a dialysate. Further, when a new feature of this device (activated wafer electrodeionization) is combined with the performance of the device, it can be adjusted on the fly to account for changes in diet. This talk will focus on the membrane features of such a device and what is needed to make it both portable and eventually, implantable.

11e - The surface properties and biological functionality of diamond coatings and membranes

Roger Narayan (UNC/NCSU Biomedical Engineering)*

The terms nanocrystalline diamond and ultrananocrystalline diamond used to describe coatings that contain small-scale diamond grains, which are circumscribed by high energy grain boundaries. Owing to the short distance between carbon atoms (1.54 Å), diamond exhibits the highest atomic density. Diamond is resistant to corrosion under physiologic conditions; corrosion only occurs in an oxygen atmosphere at temperatures over 800 °C. Due to its unique mechanical, chemical, optical, and biological properties, diamond has been considered for use as an electrode for neural prostheses and as a coating for retinal prostheses, heart valves, as well as other medical devices. The hemolysis, non-activated partial thromboplastin time, platelet adhesion/aggregation, and platelet activation of nanocrystalline diamond coatings on silicon wafers was evaluated [https://doi.org/10.1002/jbm.a.35872]. The coatings were shown to not produce a hemolytic response. In addition, the number and morphology of platelets on the coatings were similar to those on high-density polyethylene, which served as a negative control. Furthermore, few or no thrombi or platelet aggregates were noted on the coatings. These results indicate that nanocrystalline diamond coatings exhibit favorable blood-surface interactions for use in blood-contacting medical devices. In another study, free-standing nano-porous ultrananocrystalline membranes containing pore sizes of 100 nm or 400 nm were created using reactive ion etching [https://doi.org/10.1098/rsfs.2017.0063]. Attachment of SK-N-SH neuroblastoma endothelial cells on both porous regions and solid surfaces of the membranes was noted. More recently, the surface of nitrogen-incorporated ultrananocrystalline diamond coatings was modified using a fluorine plasma treatment; the fluorination process altered the chemistry, wettability (as determined by contact angle measurements), and zeta potential of the coated surface [https://doi.org/10.1016/j.matlet.2021.129823].

11f - Implantable nanofluidic membrane technology platforms for controlled drug delivery

Alessandro Grattoni (Houston Methodist Research Institute)*

Unprecedented development in controlled delivery systems have allowed for mimicking of endocrine glands responsive to biological cues for the treatment of hormonal and metabolic disorders. In this context, micro and nanofluidic membrane systems represent unique platforms that can be exploited to recapitulate organ functions or enable superior control of drug administration. By implementing cutting-edge silicon microfabrication techniques, we created nanofluidic membranes with dense arrays of monodisperse nanochannels as small as 2.5 nm. Two implantable platform technologies were developed based on these membranes, which leverage fluid physics at the nanoscale: a zero-order drug delivery system and and electrostatically gated membrane for remotely controlled and tunable administration of therapeutics. These implantable technologies were used for the sustained controlled delivery of therapeutics for the prevention and treatment of chronic pathologies. Specifically, we demonstrated their long-term function in the context of HIV pre-exposure prophylaxis (PrEP) in non-human primates, intratumoral cancer immunotherapeutics in syngeneic orthotopic models of triple negative breast cancer, and muscle wasting in microgravity. Further using the electrostatic-gated platform, we achieved reproducible drug release modulation via bluetooth remote control as a telemedical approach for the management of hypertension. This presentation will provide an overview of our platform technologies with focus on the nanoscale transport phenomena and their translation for medical applications on-Earth and in Space.

12a - A Computational and Experimental Test Bed for Prediction of RO Module Fouling

Daniel J Miller (Lawrence Berkeley National Laboratory); Sergi Molins (Lawrence Berkeley National Laboratory); Sarah M Dischinger (Lawrence Berkeley National Laboratory)* ; Mostafa Nassr (UT Austin)* ; Nicolas Spycher (LBNL); Nils Tilton (Colorado School o

In large-scale membrane systems, fouling is typically addressed with a reactive, rather than proactive, approach (e.g., cleaning once flux drops below a threshold value). Changes in process conditions, such as variations in feed water quality, hydrodynamics, and recovery frustrate efforts to control fouling while maximizing system efficiency. This project seeks to develop computational models simulating feed flows in spiral-wound RO modules, including the impacts of channel geometry, feed spacers, and inorganic scaling. We leverage previous Department of Energy investments in geochemistry models developed to understand reactive transport processes occurring in subsurface brine flows. These geochemistry models are informed by computational fluid dynamics models that describe flow around feed spacers. The models seek to capture coupled reaction-transport phenomena occurring near, on, and through an RO membrane, including scale growth and unsteady flow effects resulting from feed spacer geometry. Computational models were able to predict the water permeate flux for commercial RO membranes with and without feed spacers. Models were validated against manufacturer specification and measurements made on bench-scale crossflow experiments for commercial membranes and feed spacers. Models described scaling of commercial membranes by feed waters containing single salts and complex salt mixtures. Experimental crossflow fouling experiments in both constant flux and constant transmembrane pressure modes were compared to model predictions and to each other. Commercial and 3D-printed feed spacers were studied in crossflow fouling experiments and modeled using computational fluid dynamics techniques. Analytical techniques were developed for post-mortem membrane autopsy, including methods for characterizing mineral composition and the areal extent of fouling. Future work will extend predictive models to the effects of organic fouling and biofouling.

12b - Reduced-order models of concentration polarization in RO systems with spacers

Nils Tilton (Colorado School of Mines)* ; Jacob Johnston (Colorado School of Mines); Pedram Bigdelou (Colorado School of Mines); Sergi Molins (LBNL); Nicolas Spycher (LBNL); Sarah M Dischinger (Lawrence Berkeley National Laboratory); Mostafa Nassr (UT Au

Reverse osmosis (RO) systems depend on mass transport phenomena that are difficult to observe experimentally because they occur between tightly-packed, optically opaque, membrane sheets. This has motivated considerable computational fluid dynamics (CFD) studies that simulate fluid flow and mass transport using the Navier-Stokes and advection-diffusion equations. Such simulations are computationally expensive, particularly when the RO system includes feed spacers. Thus motivated, this NAWI project explores the development of two reduced-order models that circumvent the simulation of the Navier-Stokes equations. The first replaces the Navier-Stokes equations with an analytical expression for a feed velocity field that includes counter-rotating vortices. The vortices mimic the impact of spacers by generating alternating regions of solute accumulation and depletion on the membrane surface. The second approach avoids the Navier Stokes equations by introducing a dispersion model in the advection-diffusion equation. We compare the models in terms of computational savings and ability to reproduce the results of CFD analyses and bench-scale experiments. Overall, the dispersion model is computationally more efficient, and successfully reproduces system-level experimental measurements. Meanwhile, the vortical model is computationally more expensive than the dispersion model (though still much cheaper than CFD), but more accurately approximates CFD results for local variations of concentration along the membrane surface.

12c - A membrane fouling simulator system and omics pipeline for in operando biofouling characterization in desalination facilities

Manish Kumar (University of Texas at Austin)* ; Weiliang Bai (University of Texas at Austin); Frank Löffler (University of Tennessee, Knoxville); Mircea Podar (Oak Ridge National Labs); Amanda May (University of Tennessee, Knoxville); Lauren Stadler (Rice

Biofilms and biofouling remain highly problematic in many water systems and are especially detrimental in membrane filtration systems. Biofouling is generally regarded as the Achilles heel of reverse osmosis (RO) membrane application for desalination and water reuse due to two reasons: i) spiral wound RO membranes cannot be backwashed frequently like ultrafiltration and microfiltration hollow fiber membranes, and ii) free chlorine, which can be applied in many water treatment scenarios for biocontrol, cannot be used for polyamide RO membranes that are highly sensitive to chlorine. To explore biofilm formation, a new NAWI project collected biofilms from RO modules and pretreatment systems from the two largest seawater desalination facilities in the US. Standard operating procedures for the application of omics tools (e.g., metagenomics, metabolomics) are being developed to allow monitoring of biofilm dynamics and capture changes in the microbial community during the transition from benign to detrimental biofilm. To avoid interference with plant operations during sampling, a membrane fouling simulator system was installed on site. The initial findings suggest that an online biofilm sampling system combined with omics tools analysis could be an excellent approach to indirectly monitor biofouling in the RO modules, and implement ecologically informed control measures (e.g., phage therapy) to reduce downtime, lower the energy demand, and prolong the lifetime of the RO modules.

12d - Enhanced Wastewater Reclamation with Carbon Molecular Sieves

Haley D White (Georgia Institute of Technology)* ; Ryan Lively (Georgia Tech)

The global water crisis demands robust technologies to support reuse of impaired water sources. Water reuse has been limited by the poor ability of available technology to remove certain small, neutral molecules. We hypothesize that such emerging contaminants can be removed with pure carbon molecular sieve (CMS) membranes, as they have already exhibited size and shape selectivity in gaseous and organic separations due to the presence of permanent, rigid, and molecular sieving pores. Differences in the application of the sorption-diffusion model for transport of gaseous versus aqueous separations in CMS materials will be discussed. Key learnings from this discussion will be applied to a newly developed CMS material, derived from a highly selective aromatic polyamide. Nitrogen physisorption measurements of CMS indicate an ultramicropore size range of 5.5 to 6.5 Å and a micropore range of 10.3 to 10.6 Å. Structural characteristics are further investigated with x-ray diffraction and Raman spectroscopy. 1HNMR, mass spectrometry, and x-ray photoelectron spectroscopy are used to hypothesize degradation mechanisms leading to this CMS structure. While challenges to the development and implementation of such materials as membranes remain, highly selective materials are of paramount importance to sustainable membrane-based separation processes of the future.

12e - Highly Precise Ion Separations via Polymeric Membranes with Host-Guest Chemistry

Ryan DuChanois (Yale University)*; Mohammad Heiranian (Yale University); Jason Yang (Yale University); Cassandra Porter (Yale University); Xuan Zhang (Nanjing University of Science & Technology); Rafael Verduzco (Rice University); Menachem Elimelech (Yale

Water and energy technologies that separate ionic species of nearly identical size and equal charge are essential to the transition to a circular economy and clean energy future. However, state-of-the-art polymeric membranes cannot attain the precise ion separations needed to meet emerging objectives, such as recovering valuable metals from wastewater. Membranes with reversible coordination interactions with target ions, as seen with biological ion channels, may be required to obtain selectivity between species of similar physicochemical characteristics. In this study, we systematically assess the permeability of five divalent cations of similar size — Cu^{2+}, Ni^{2+}, Zn^{2+}, Co^{2+}, and Mg^{2+} — through metal-binding polymer membranes in diffusion dialysis. Specifically, we synthesize a poly[(N,N'-dicarboxymethyl) allylamine] (PDCMAA) metal-binding agent via carboxymethylation of poly(allylamine) (PAH), and then prepare membranes via layer-by-layer deposition of PAH and PDCMAA onto anodic aluminum oxide substrates. We report that metals with higher binding energy to the polymer, which we determine via density functional theory calculations, selectively pass through the membrane without evident reduction in permeability. In contrast, passage of weaker binding species is significantly reduced because stronger binding species preclude their complexation with the polymer. This permeability reduction is roughly proportional to the difference in metal binding energy to membrane binding sites and enhances selectivity for strong-binding species markedly (e.g., $Cu^{2+}/Mg^{2+} > 50$). Furthermore, data projections suggest that ultra-thin polymer films (< 1 nm) provide considerably higher selectivity ($Cu^{2+}/Co^{2+} > 30$) compared to 10-µm thick membranes ($Cu^{2+}/Co^{2+} < 6$). Our findings indicate that ultra-thin polymeric membranes with tailored binding sites can produce highly precise separations that may enable recovery of high-value metals such as Cu^{2+} from mining wastewater.

12f - Pore Functionalized Catalytic and Responsive Membranes for Water remediation

Dibakar Bhattacharyya (U. Kentucky)* ; Francisco Leniz (University of Kentucky); Rollie Mills (U. Kentucky); Ronald Vogler (University of Kentucky); Lindell Ormsbee (University of Kentucky)

The development of catalytic and responsive, multifunctional materials and membranes for water applications requires a high level of control of both the characteristics of the base polymeric support layer, as well as, its corresponding pore configurations. Membrane processes provide a highly flexible separation technique for water treatment, toxic materials degradation, and materials recovery. Microfiltration membranes are primarily used for the removal of bacteria and particles, but by appropriate pore functionalization with polymeric charge groups or integration of nanoparticles of these MF membranes (termed Functionalized Membranes) one can use these as high capacity metal capture membranes to catalysis and pH/temperature flux modulations with separation selectivity. Many current treatments for the reclamation of contaminated water sources are chemical-intensive, energy-intensive, and/or require post-treatment due to unwanted by-product formation. We demonstrate that through the integration of nanostructured materials within pore-functionalized synthetic membrane platforms (such as, PVDF membrane pores functionalized with poly-acrylic acid, PAA and/or PNIPAM), we are able to conduct environmentally important reactions for toxic organic degradation (PCB, TCE, etc.) and selective toxic material sorption from water without the addition of expensive or harmful chemicals. In addition, we have fabricated GO-PAA composite membranes by layer-by-layer assembly with an aim to integrate advanced sulfate radical-based oxidation reactions for efficient removal of chlorinated organics from water. The presentation will include: (a) combined pH and temperature responsive properties of pore functionalized membranes and in-situ catalyst synthesis, (b) enhanced toxic organic sorption and reaction rates of toxic organics through thermal modulation, and (c) GO-PAA composites for water and catalysis related applications. This research has been supported by NIH-NIEHS-SRP.

13a - Spatially Controlled Permeability and Stiffness in Photopatterned Glass-Rubber and Rubber-Rubber Two-Stage Reactive Polymer Films

Adrienne K Blevins (CU Boulder)* ; Michelle Lehmann (University of Tennessee); Mengyuan Wang (Oak Ridge National Lab); Lewis Cox (Montana State University); LEIQING HU (University at Buffalo); Jasper Drisko (NIST); Haiqing Lin (The State University of New

Heterogenous polymeric materials, such as polymer nanocomposites, are commonly researched for many transport-based applications. These multi-component materials are complex systems which often encounter a wide variety of fabrication issues, limiting control over the material structure. Structure-property relationships strongly impact key performance parameters such as gas permeability and mechanical properties. As such, controllably structured polymeric materials are attractive. In this work we utilize Two Stage Reactive Polymer (TSRP) networks to generate films with domains of different degrees of crosslinking. This results in chemically similar domains with different permeability and modulus values. Using photopatterning, we can generate a wide variety of different domain geometries with strong interfaces. In this talk, I will discuss the formulation and patterning of two different TSRP systems. The impacts of both pattern geometry and volume fraction on transport and mechanical properties of the films are determined. The results show that more superior combination of mechanical properties and gas permeability (CO_2 and N_2), as compared to homogenous component films, can be achieved using this approach. This method can be beneficial for a variety of applications where balancing mechanical properties and transport properties is often a challenge.

13b - Characterization of physical aging-induced evolution of CMS membrane using a dual-mode sorption and transport model

Zhongyun Liu (Georgia Institute of technology)* ; William Koros (Georgia Institute of Technology)

Abstract

Scalable carbon molecular sieve (CMS) membranes provide outstanding separation performance for many industrially important gas pairs. Freshly-made CMS membranes exhibit physical aging reflected by permeance reduction and selectivity increases. CMS membranes derived from 6FDA-BPDA/DAM (1:1) polyimide exhibit complex morphologies that can be analyzed in terms of distributed molecular sieving Langmuir domains within a disordered continuous phase. Characterization of structural evolution apparent in the two discrete environments during physical aging provides a new tool to understand CMS membranes. In this study, CMS hollow fibers derived from 6FDA-BPDA/DAM (1:1) polyimide, prepared via pyrolysis at 550 °C under inert argon atmosphere, were considered for pure components and a 50/50 C_3H_6/C_3H_8 feed mixture. A combined dual-mode sorption and transport model is used to deconvolute sorption and transport factors within the material. The results show clearly that the CMS aging is primarily "diffusion related". Moreover, most diffusion related changes were apparent in the disordered continuous phase, rather than in the molecular sieving Langmuir phase. Possible implications of such trends on strategies to control physical aging and even use it as a tool to tune performance will be considered.

13c - Elucidating the role of side chain length and dispersity in ROMP polymers with pore-generating side chains for gas separations

Sharon Lin (Massachusetts Institute of Technology)* ; Francesco Maria Benedetti (Massachusetts Institute of Technology); You-Chi Wu (Massachusetts Institute of Technology); Timothy Swager (Massachusetts Institute of Technology); Zachary Smith (MIT)

Recently, CF3-ROMP and OMe-ROMP polymers were synthesized via ring opening metathesis polymerization (ROMP) for gas separation applications. These two polymers exhibited unprecedented CO_2 plasticization resistance, and CF3-ROMP further showed ultrahigh CO_2 permeability that exceeds those of most polymers in literature. A unique feature of CF3-ROMP and OMe-ROMP is that they possess pore-generating side chains that vary in length. In this study, we report gas transport properties on OMe-ROMP with uniform side chain lengths ranging from n = 2–5 repeat units to elucidate the role of side chain length and dispersity on gas transport properties. It was found that increasing the side chain length led to increased BET surface area and gas permeability, with minimal loss in selectivity. In addition, increased plasticization resistance was observed, which can be attributed to the increased interchain rigidity afforded by longer side chains. Controlling the side chain length was revealed to be an effective strategy to rationally optimize the performance of ROMP polymers for CO_2-based gas separations.

13d - Greener preparation of defect-free asymmetric gas separation membranes with dihydrolevoglucosenone (CyreneTM) as an alternative polar aprotic solvent

Alexander Bridge (The University of Texas at Austin)* ; Benjamin Pedretti (The University of Texas at Austin); Joan Brennecke (The University of Texas at Austin); Benny Freeman (The University of Texas at Austin)

Membrane-based gas separation processes favor technology that offers high productivity (flux) without sacrificing product purity. To meet these criteria, nonsolvent-induced phase separation (NIPS) is used to prepare membranes with an asymmetric morphology. NIPS casting solution formulations for current gas separation membranes are limited to a small group of glassy polymers and, importantly, contain toxic polar aprotic solvents. Growing restrictions on the industrial use of polar aprotic solvents necessitates the development of more benign formulations that do not compromise membrane performance. This work reports, for the first time, the development of high-flux and defect-free asymmetric polysulfone (PSf) gas separation membranes that use dihydrolevoglucosenone (Cyrene), a bioderived dipolar aprotic green solvent, as the majority formulation component. Optimized formulations and casting conditions produce membranes with hydrogen permeances in excess of 100 gas permeance units (GPU) and selectivities at or above those exhibited by dense PSf films produced via evaporative casting. Membrane performance improved with shorter dry step times and increased loadings of Cyrene relative to the volatile solvent, tetrahydrofuran (THF), in the casting solution. Unique interactions between Cyrene, the nonsolvent (water), and PSf help suppress support layer macrovoid formation, providing insight on the relationship between support layer morphology control and the performance of defect-free asymmetric membranes.

13e - ZIF-21 Membranes for Ammonia Separation

Qiushi Wei (Colorado School of Mines); Jolie Lucero (Colorado School of Mines); James Crawford (Colorado School of Mines); J. Douglas Way (CSM); Colin Wolden (Colorado School of Mines); Moises A Carreon (Colorado School of Mines)*

LTA type zeolitic imidazolate framework membranes denoted as ZIF-21 were synthesized and evaluated for the separation of a commodity chemical, ammonia, from nitrogen and hydrogen. Membranes were grown using secondary seeded growth. Ammonia permeances were as high as 1727 GPU, with NH3/N2, and NH3/H2 ideal selectivities of 35, and 12 respectively. ZIF-21 membranes were also evaluated for gas mixtures composed of 13.8:86.2 NH3/N2 and NH3/H2, with mixture separation selectivities as high as 7.6 for NH3/N2. The separation of these mixtures was kinetically and thermodynamically driven. Specifically, for both gas pairs the polar-polar interaction between ammonia and the linker polar channels, led to the preferential adsorption of ammonia, and therefore favored thermodynamically the separation of ammonia from nitrogen and hydrogen. In addition, for NH3/N2 gas pair, the separation was kinetically driven, via differences in diffusivities. For NH3/H2 gas pair, due to the faster hydrogen diffusivity, the kinetic contribution was a strong competing separation mechanism.

13f - Performance of Gas Separation Hollow Fiber Membrane Modules Fabricated from Fiber Tows

Lili Sun (University of Toledo); Grigorios Panagakos (Carnegie Mellon University); Glenn Lipscomb (University of Toledo, Toledo)*

Commercial hollow fiber membrane manufacturing processes commonly use hollow fiber spinning to produce multiple fibers from one spinneret head. These fibers move together through the manufacturing process and are collected on a spool at the end of the process. This group of fibers is referred to as a fiber tow.

Modules are fabricated from fiber tows by wrapping or weaving tows to form a cylindrical bundle. Tubsesheets are formed at one or both ends of the bundle to create a module that is placed in a case for use in a separation process.

The use of fiber tows introduces two length scales in the fiber bundle: 1) the nominal spacing between fibers in the tow and 2) the nominal spacing between tows. The existence of two length scales for shell-side axial flow channels exacerbates the residence time distribution that exists in the shell-side flow which can be detrimental to performance.

The changes in residence time distribution associated with bundles of fiber tows is evaluated for a range of intra and inter-tow fiber packings. The effect increases as the packing fraction difference increases.

The effect of the residence distribution on module performance is evaluated for operating conditions anticipated for carbon dioxide capture applications from fossil-fuel fired power plants. The bimodal residence time distribution is detrimental to performance, especially as the stage cut increases. The effect on performance is correlated with the standard deviation of the residence distribution and the separation of the bimodal peaks.

14a - Electron tomography for the characterization of membranes

Enrique Gomez (Penn State); Michael Geitner (The Pennsylvania State University)*

Advances in electron microscopy continue to push the limit of what is visible at the atomic and molecular scale. An emerging opportunity is the combination of multiple techniques within the electron microscope to reveal new types of information that is otherwise not accessible. For example, we have combined the focused ion beam with scanning electron microscopy through serial sectioning to reconstruct a 3D representation of ultrafiltration membranes using for virus removal from biopharmaceutical streams. In addition, we have combined energy-filtered transmission electron microscopy with electron tomography from scanning transmission electron microscopy images to map the variation in density of polyamide films used in reverse osmosis membranes. Quantitative analysis of imaging products are key to extract mechanistic details that govern water transport and separations.

14b - Compositional analysis of polyamide membranes via 13C MAS NMR spectral editing

Christopher M Stafford (NIST)*; Ryan Nieuwendaal (NIST); Peter Beaucage (NIST); Jefferey Wilbur (Dupont)

Establishing robust relationships between the chemical structure, water/ion transport, and water filtration performance in the polyamide selective layer of RO/NF membranes remain elusive. Previous literature results from solid-state NMR have shown that T1?H is an important metric for polyamide segmental dynamics upon swelling, and that 15N {1H} CPMAS with DNP can be used to discriminate amine and amide populations in the bulk and on the surface. Here, we use 13C MAS NMR measurements to quantify purity and composition of a series of polyamide membranes that exhibit variations in performance. We show that 1H-T1 based 13C CPMAS spectral editing is useful for discriminating polyamide 13C resonances from those of polysulfone and other impurities, which is important for quality analysis of the membrane harvesting procedures. We then show that the carbonyl 13C intensity is useful for quantifying composition in these samples, allowing acid:amine monomer ratios to be determined. Structural characterization of these samples via grazing incidence wide angle x-ray scattering (GI-WAXS) also provides some insight into the packing efficiency of this series of polyamide membranes, and we correlated the packing efficiency to the compositional and membrane performance data.

14c - Using microfluidic interferometry to visualize diffusive solute gradients within polymer membranes

Varun H Hegde (University of California Santa Barbara)* ; Todd Squires (University of California Santa Barbara)

The measurement of solute permeation under known concentration drops has been pivotal in characterizing membrane materials, but generally can not elucidate concentration profiles within membranes. Here we introduce microfluidic interferometry, an in-situ Fabry-Perot interferometric technique, as a means of visualizing the spatio-temporal evolution of solute concentration gradients within polymer gels. By tracking Fringes of Equal Chromatic Order (FECO), a light interference pattern, in a gel within a microfluidic channel, the swelling of the polymer network and the evolution of solute concentration gradients can be captured simultaneously. While most characterization techniques determine an average mutual diffusivity, this dynamic technique is capable of determining the local concentration dependent mutual diffusivity and partition coefficient. Furthermore, by visualizing the development of diffusive profiles within membranes during performance, solute rejection and adsorption can be investigated.

We demonstrate the versatility of this technique by studying the diffusion of Sodium chloride, Boric acid, and Methanol in Polyethylene glycol diacrylate(PEGDA) hydrogels under varying pre-polymerization conditions. This versatile technique is optimized to characterize novel membrane materials that can be cross-linked or polymerized within microfluidic devices. We envision a microfluidic device that can incorporate external samples which will allow for the characterization of a wide range of solutes, structured and unstructured membranes, under a host of operating conditions.

14d - Elucidating the Fundamental Mechanisms of CO_2 Facilitated Transport in Amine-functionalized Polymeric Membranes Using Operando Spectroscopy

Casey O'Brien (University of Notre Dame)* ; Sarah Pate (University of Notre Dame); Hui Xu (University of Notre Dame)

Reactive amines in polymeric facilitated transport membranes (FTMs) facilitate CO_2 transport across the membrane to give rapid and selective CO_2 separation from mixed gas streams such as natural gas and flue gas. However, the fundamental mechanisms by which the amine carriers facilitate CO_2 transport are not well understood. For example, it is not clear whether the amine group catalyzes the transformation of CO_2 into bicarbonate species, which are mobile within the polymer matrix, or if CO_2 hops from one amine group to another as a carbamate species, which is chemically attached to the polymer chain. It is also possible that CO_2 moves across FTMs by some other unknown mechanism.

In this talk, I will describe a unique operando spectroscopy methodology that we have developed to elucidate the fundamental mechanisms of amine-facilitated CO_2 transport. Specifically, we have developed two complementary tools—one that utilizes transmission FTIR spectroscopy and one that utilizes Surface Enhanced Raman Spectroscopy (SERS)—that are capable of identifying CO_2 transport intermediates formed in the membrane and correlating their formation to CO_2 permeation rates under realistic operating conditions. I will show that both tools indicate that CO_2 moves across polyvinylamine FTMs primarily as carbamate species. I will also demonstrate, for the first time, direct observation of the carrier saturation phenomenon. The mechanistic insights obtained from this new operando methodology will enable a more rational design of FTMs than conventional membrane testing with ex-situ characterization approaches.

14e - Practical limits of the quartz crystal microbalance for elucidating membrane phenomena (V)

Jon Arrizabalaga (Surphase); Iliane Rafaniello (SURPHASE); Thomas Schäfer (Polymat, University of the Basque Country)*

Determining the partitioning of salts into polyamide (PA) reverse osmosis membranes is indispensable for optimizing their salt retention. Recently, quartz-crystal microbalance with dissipation monitoring (QCM-D) has been explored for this purpose. While this piezoelectrical technique has in theory a sensitivity in the range of pg/cm2, its practical sensitivity is orders of magnitude lower [1,2] and it is furthermore highly sensitive to bulk changes (e.g., viscosity). The utility of QCM-D for quantifying partitioning of salt into PA membranes was therefore systematically investigated. The experiments were complemented by multiparameter surface plasmon resonance (MP-SPR) as an optical surface-sensitive technique.

During this study, sodium chloride and sucrose (as control) solutions were contacted with PA as well as gold (reference surface) and PEBAX (as control polymer). The results obtained revealed that it could not be discriminated between a supposed salt partitioning and the sensors' response to a bulk change. In fact, the response of the viscosity-sensitive sensor was found to be directly correlated with the change dynamic viscosity of the bulk solutions. MP-SPR data corroborated this observation.

It was proven that while QCM-D is in practice a very convenient and easy method to study polymer-solute interactions, great care must be taken during the experimental design and interpretation of the sensor signal such as not to draw erroneous conclusions. The benefits of combining complementary surface-sensitive techniques will be shown such as to minimize the risk of pitfalls.

[1] Chem. Commun., 2012, 48, 10087-10089

[2] Chem. Sci., 2017,8, 7038-7046

14f - Particle Remobilization in Filtration Membranes during Flow Interruption

Haichao Wu (University of Colorado Boulder)* ; Daniel Schwartz (University of Colorado Boulder)

Anomalously high levels of particles are often observed to pass through filtration membranes upon resumption of flow following interruption. However, it has been challenging to obtain direct information about the underlying mechanisms. Here, we employed a highly multiplexed single-particle tracking approach, which enabled the visualization of particle transport in microfiltration membranes under flow conditions that included stoppage intervals and therefore provided direct evidence for particle remobilization during flow stoppage. Our results suggest that during flow interruption, a sub-population of trapped/retained particles detach from their retention sites and may escape to less trapped areas through Brownian motion within the pore space. Particles that do not diffuse far enough during flow stoppage generally become trapped again upon the resumption of flow. As a result, the fraction of particles that are remobilized increases with stoppage time, since longer periods of Brownian motion permit diffusing particles to enter less confined regions of the membrane. This mechanistic information about particle transport remobilization will permit the design of filtration flow profiles that meet customized requirements for particle retention during complex filtration cycles.

15a - Membrane-based Ventilation Energy Recovery: Current Industry Perspectives (V)

Ryan Huizing (CORE Energy Recovery Solutions)*

Most modern buildings use mechanical ventilation to extract exhaust air and supply fresh air to the indoor environment to ensure good indoor air quality. More recently increased ventilation to remove contaminated indoor air and supply more fresh outdoor air has been used to combat the spread of COVID-19 in indoor spaces. Dedicated outdoor air systems providing 100% outdoor air with no recirculation and increased air change rates have been recommended. Ventilation, however, has an energy penalty -- outdoor air must be heated or cooled, and humidified or dehumidified in order to maintain occupant comfort.

In order to address this issue, HVAC engineers rely on energy recovery ventilation to transport heat and moisture between incoming and exhaust air streams in building ventilation systems. Energy recovery ventilation (ERV) minimizes the energy consumption associated with ventilation. Over the last 10 years, this has increasingly been accomplished using polymeric membrane-based exchangers. Over this time, CORE Energy Recovery Solutions (formerly dPoint Technologies) has fielded hundreds of thousands of ventilation membrane exchangers and produced several millions of square meters of polymeric water vapour transport membranes.

This presentation provides an overview of ventilation energy recovery using membrane-based enthalpy exchangers. It will discuss the unique membrane requirements for ERV applications, including high water vapour permeability, water/VOC selectivity under humid conditions, durability, and cost. We will summarize our ongoing membrane and exchanger research and development activities at CORE and provide perspectives on areas for future research needs.

15b - A comprehensive overview of Liquid-to-Air Membrane Energy Exchanger (LAMEE) for building HVAC application

Gurubalan Annadurai (University of Saskatchewan)* ; Carey J. Simonson (University of Saskatchewan)

Air conditioning (AC) is the process of treating the air to meet the requirements of a conditioned space by controlling its temperature, humidity, cleanliness, and distribution. It is predicted that AC systems will consume the maximum electricity in buildings and will account for 16% of the global electricity demand by 2050. More than 90% of the conventional AC systems (CACS) are vapor compression refrigeration type. CACS is energy inefficient because of the overcooling and the subsequent reheating of air for its humidity control. A liquid desiccant system is a prospective alternative to CACS for humidity control. Its advantages are accurate humidity control, better indoor air quality (IAQ), and a higher coefficient of performance. It mainly consists of a dehumidifier and regenerator which are used to control the air supply conditions (temperature and humidity) and regenerate the desiccant (for continuous operation) respectively. The dehumidifiers and regenerators are broadly classified as packed bed and membrane types. In the former, air carries tiny droplets of desiccant due to their direct contact. This carryover deteriorates IAQ and corrodes the air handling equipment. Membrane type uses a microporous membrane that separates the air and desiccant and allows only the water vapor through its pores. This avoids the desiccant carryover and hence, the membrane types are attractive for AC application. Membrane type dehumidifiers and regenerators are commonly referred to as Liquid-to-Air Membrane Energy Exchanger (LAMEE). LAMEEs have been researched for more than two decades. The design, steady and transient performances of LAMEE and its application in air conditioning were studied in detail. The energy and economic performances of run-around energy exchanger (which consists of two LAMEEs) were investigated. The selection and properties of the membrane and desiccant solutions were studied. The present study gives a summary of these research works.

15c - Liquid Desiccant Air Conditioning Using Selectively Permeable Membranes

Daniel Betts (Blue Frontier, LLC); Matthew Graham (Blue Frontier, LLC): Matt Tilghman (Blue Frontier, LLC)*

Liquid desiccants can be used to dehumidify air in HVAC applications. One such application uses a series of plates that form channels where air flows. The initial portion of the walls of these channels have liquid desiccant, in high concentration, flowing, and a selectively porous membrane separates the air flow from the liquid desiccant. The selectively porous membrane allows the passage of water vapor through it, but not liquids, which permits the exchange of water vapor between the air flow and the liquid desiccant on the other side of the membrane, creating dehumidified air. This dehumidified air can then be subjected to indirect evaporative cooling. This process can produce air conditioning with much lower energy requirements than conventional direct expansion systems. However, in order to make the system compact and reliable, it relies on membranes. A modeling study has been conducted to evaluate potential designs of this membrane-enabled liquid desiccant air conditioning system. In particular, sensitivity of efficiency to membrane vapor permeability is analyzed and presented.

15d - Efficient Dehumidification Using Membranes in the CCL Process

David E. Claridge (Texas A&M University)* ; Charles Culp (Texas A&M University)

This presentation will cover a basic model for analyzing the performance of the Claridge-Culp-Liu dehumidification process. The fundamental process efficiency limit for dehumidification is close to COPCarnot, but for the eight dehumidification cases examined, the limiting or ideal energy use required is 26% to 56% that of a Carnot condensing system. This presentation will emphasize the impact of membrane area, water vapor permeance, non-zero air permeance, and non-zero system air pressure drop on the dehumidification efficiency of the system. The impact of non-ideal compressors, vacuum pumps, and condensers will also be discussed. The performance of a "conservative" membrane system based on the use of existing components is computed for eight specific conditions along with that of a system that assumes expected component performance after additional future component development. The "conservative" membrane system would use 36% to 66% as much energy as a system with a COP=7 chiller to produce the same dehumidification for the eight cases examined while the improved system would use 15% to 40% the energy of a system with a COP=7 chiller. In addition to the significant energy reduction over conventional technology, the membrane system offers the advantages of: 1) no HFC refrigerant use; 2) direct isothermal control over humidity ratio setpoint; 3) maximum capacity occurs at design conditions; and 4) the system generates pure water extracted from air as a by-product.

15e - High Efficiency, Water Vapor-Selective, Active Membrane Energy Exchanger for Air Conditioning

Andrew Fix (Purdue University)* ; David Warsinger (Purdue); James Braun (Purdue University)

Water vapor-selective membranes, which allow water vapor transport but block other gasses like N2 and O2, can enable a highly selective mechanical separation of humidity out of other gasses. One prominent application for these membranes includes heating, ventilation, and air conditioning (HVAC) systems. As much as 70% of the energy used in conditioning outdoor air could stem just from the condensation dehumidification process, which is required to maintain both occupant comfort and health. Our research group has proposed a novel membrane-based HVAC system, referred to as the Active Membrane Energy Exchanger, which uses water-vapor selective membranes to dehumidify air while simultaneously cooling the air using conventional cooling technologies. This device is a hybrid of flat sheet membrane modules and counter flow tube heat exchangers. This approach is the first of its kind to deliberately target non-isothermal membrane dehumidification in a field that has rigidly studied membrane dehumidification as an isothermal process. Preliminary modeling results suggest that this system can attain upwards of 60-70% energy savings and 20-50% second law efficiency in air conditioning operations by avoiding the large energy requirements associated with phase change dehumidification and by exploiting other thermodynamic benefits. While some passive membrane HVAC systems for energy recovery have made their way into the market, active dehumidification and cooling technologies have not. In this presentation, we will review the fundamentals of the Active Membrane Energy Exchanger and provide an overview of the thermodynamic benefits of such a system. Furthermore, we will present details regarding our prototype system design and preliminary experimental results. We also aim to address the key technical challenges to help foster an understanding of how the membrane science community can play a critical role in advancing these technologies.

15f - Multifunctional membranes for managing moisture and heat in buildings

Derek Stein (Techstyle Materials, Inc.)* ; Ravi Kishore (National Renewable Energy Laboratory); Xuejian Lyu (Techstyle Materials, Inc.); Zoe Kaufman (National Renewable Energy Laboratory)

Buildings account for nearly 40% of the primary energy consumed in the United States. A fundamental problem is that increasing the insulation (R-value) and airtightness of a building to improve its thermal efficiency typically also increases its susceptibility to moisture damage. There is an urgent need for materials that simultaneously increase the energy efficiency, durability, and interior air quality of building envelopes. Here we introduce a multi-functional membrane concept for passively managing moisture and heat in buildings envelopes. The membrane exhibits directional vapor permeance, which increases the durability of a building by reducing the peak moisture level inside the wall assembly and by creating a pumping effect that actively removes moisture from the wall cavity. The membrane also adsorbs and redistributes water within its structure in ways that regulate the relative humidity in the enclosure, and this improves indoor air quality and occupant comfort. Finally, the membrane reduces temperature fluctuations and energy use for heating and cooling by storing and releasing large quantities of latent heat as water vapor respectively evaporates and adsorbs within its structure. We will describe the heterostructure that defines the multifunctional membrane, explain how its two sub-layers work in concert to automatically manage flows of moisture and heat, and present the results of numerical calculations of the membrane's impact on the moisture and energy performance of buildings.

16a - Some industrial perspectives on the development of novel materials for propylene/propane separations: adsorbent and membrane (V)

Jay (Junqiang) Liu (The Dow Chemical Company)*; Derrick Flick (Dow Inc.); Abhishek Roy (DOW); Barry Fish (The Dow Chemical Company); Gerard Rogers (Dow Inc.); Surendar R Venna (Dow Inc.); Li Tang (Dow Inc.)

Propylene is one of the major basic feedstocks for the chemical industry. The separation step contributes to a significant portion of the production cost. Pressure swing adsorption (PSA) and membrane systems have potential to reduce this energy consumption. The key for successful PSA or membrane systems will be advanced materials, such as Metal Organic Framework, Zeolite, Carbon Molecular Sieve (CMS). We have evaluated several types of materials, including zeolite 4A, commercial CMS, and Dow in-house made CMS, for propylene/propane separations. This talk will discuss some industrial perspectives and learnings we found that may be useful to bridge the industry - academic researches. In order to be used in industrial streams, adsorbent and membrane materials have to meet several key criteria: 1) a broad range of operating temperature; 2) not impacted by impurity species, such as water and methanol; 3) inert surface to avoid reactive chem issues; 4) robust to run multiple years, etc.

Our team has developed CMS pellet adsorbent prototype and tested on bench-scale PSA. There are three key learnings: 1) raw material cost is only a portion of CMS adsorbent production; 2) the cycle time is long for industrial PSA process and the footprint of adsorbent bed is large; 3) the effluent propylene product purity is limited. Compared to PSA, membrane can reduce the amount of material usage, which is a significant portion of the capital cost. The prerequisite for industrial membrane application is scalable production of thin membranes. Polyimide and PVDC based CMS showed some initial promise on this aspect. The pursuit of novel membranes is continuing in Dow to achieve several key performances: 1) permeance & selectivity; 2) mechanical properties; 3) scale up production which includes raw material, processing, and conversion.

The learning and discussion from this talk will provide some useful industrial reference when evaluating novel materials for propylene/propane separations.

16b - Organic Solvent Nanofiltration and their role in emission reduction and process intensification in the chemical/petrochemical industry

Chrisitian Schnitzer (Evonik Operations GmbH); Udo Dengel (Evonik Corporation, Inc.)* ; Kumar Abhinava (Evonik Corporation, Inc.)

Energy and resource efficient processes are more important than ever. Besides value contributions from the reduction of operating expenditures or superior product quality, emission reductions have become a bigger contribution in recent years. In membrane separation processes value for the customer is directly linked to a smart process design and to separation performance of the functional unit, the membrane module.

A good membrane module has to keep its separation performance under real process conditions. Spiral wound modules from polymeric membranes are best suited to offer good separation performance at acceptable cost. In recent years, many more applications of Organic Solvent Nanofiltration have been investigated and plants have been built utilizing the benefits of the technology.

The lecture will provide two specific examples for process solutions based on Organic Solvent Nanofiltration:

a) Comparison of OSN with distillation, evaluating economic and environmental benefits, the latter based on a full Life Cycle Assessment which has been carried out for the process.

b) Embodiment of an Organic Solvent Nanofiltration plant into existing process infrastructure in order to debottleneck the process and reduce operating expenditures.

An overview on further potential application areas will sum up the presentation. Applications for Organic Solvent Nanofiltration are manifold and can sometimes be challenging. In the absence of proper simulation tools for the technology, process design follows a stepwise, empirical approach as close as possible to the real process. These individual process developments can take quite some time and effort. The provided examples will show how rewarding it can be to work out suitable process solutions based on the technology.

16c - High Flux CO2 Selective Membranes for Renewable Natural Gas and CO2 Capture

Christine Parrish (Compact Membrane Systems); Hannah Murnen (Compact Membrane Systems)* ; Ning Shangguan (Compact Membrane System); Sudip Majumdar (Compact Membrane Systems)

Reducing carbon emissions is the key to preventing further climate change and mitigating the damage that has already been done. Compact Membrane Systems is developing a highly permeable, carbon dioxide selective membrane designed to operate at low pressure with minimal gas pretreatment. Results from two applications will be discussed: the upgrading of biogas to renewable natural gas and capture of carbon dioxide from point sources.

Biogas is a carbon neutral fuel which has a potential to reduce greenhouse gas emission significantly. According to the World Biogas Association, a full recycling of organic waste by applying anaerobic digestion for biogas production can reduce greenhouse gas emissions by 10% in 2030. CMS has developed a membrane aimed at allowing users to separate and upgrade existing biogas at all scales.

Working with the Energy Project Manager Matt Steiman at the Dickinson College Farm food waste digester, CMS deployed a small spiral wound module system using Optiperm™ biogas membranes. Raw biogas (no pretreatment) was fed directly to the membrane system and the system was able to upgrade the gas stream from 60% to 90% methane in a single stage. OptipermTM biogas technology demonstrated stability in the field for over 2 months and validated performance measured in the CMS laboratory. Using results from this pilot, CMS has modeled a full biogas system to produce pipeline quality fuel. The energy use and system size as well as economic benefits of the CMS system will be discussed.

Finally, initial results for utilizing the CMS membranes to capture carbon dioxide from flue gas, including the effect of operating conditions and initial system modeling, will be shared. This application is early in development but results thus far have shown a very high flux membrane that could result in substantially lower capture costs than the current state of the art. Modeling highlights the advantages of different membrane staging designs and operating conditions.

16d - Membrane Applications in Industrial Processes

Johannes G Wijmans (Membrane Technology and Research, Inc.)*

There are numerous opportunities for membranes to make industrial process more efficient by reducing the consumption of energy and materials. Applications currently practiced are almost exclusively gas separation applications, but pervaporation and liquid permeation applications are expected to grow in importance in the future.

These "Process Intensification" applications involve integration of the membrane process into the overall process and typically do not require that a high purity permeate or residue stream is produced. This plays into the strength of membranes as a bulk separation process. This presentation will review a number of applications and analyze why membranes are successful.

16e - Powering the Future of Energy Storage with membranes

Amit Patwardhan (Energy Exploration Technologies)*

Amit Patwardhan , Teague Egan, Nick Grundish, Richelle Lyndon, Kevin Reimund, Benny Freeman+

Energy Exploration Technologies Inc., Austin, TX

+University of Texas at Austin

Lithium is recognized to be a key raw material in the future of energy storage in the form of lithium ion batteries. Lithium is abundant but limited in high concentrations and extraction involves the use of traditional methods fraught with very low recovery and a large environmental footprint. One of the most important separations in lithium extraction is between lithium and magnesium. EnergyX has developed and scaled low-cost robust LiTASTM mixed matrix membranes (MMM) incorporating metal organic frameworks (MOF) and demonstrated a lithium over magnesium selectivity typically between 100-900 over dozens of real brines tested from existing operations and upcoming projects. Furthermore, EnergyX has identified novel application of these membranes in the production process to increase recovery by 200-400% over that achievable from traditional methods using the same bine and existing infrastructure while improving product quality and lowering production costs. Application of MOF-MMM has been further extended for use in SoLiSTM solid state electrolytes achieving conductivities exceeding 10-3 S/cm and stable performance over at least 1000 cycles tested so far. Solid state batteries are regarded to be the next generation of portable energy storage as they provide improved safety and higher energy density over current technology. EnergyX is powering the future of energy storage from mine to consumer using novel membrane applications.

16f - Separation of complex hydrocarbon mixtures by NF and RO – extending fundamental understanding to concepts for refining and petrochemicals

JR Johnston (ExxonMobile)

The fractionation of hydrocarbon mixtures by distillation forms the foundation upon which all petroleum refining is built. Membrane based separations have the potential to avoid phase changes and separate molecules based on properties other than boiling point. Therefore, in combination with distillation or as standalone separations, membranes have the potential to access molecules not separable by distillation alone and reduce the energy required to provide the products demanded by the global economy.

ExxonMobil has been performing research in the space of membrane based hydrocarbon liquid separations on and off for over 25 years. In that timeframe much has changed in the area of advanced materials for separation and in the need for energy reductions in the manufacturing space. This talk will give an overview of our work in membrane separations. We will cover the separation of wide-ranging complex hydrocarbon streams from vacuum tower bottoms to mixed xylenes with membranes covering the spectrum of material types. We will also discuss our progression in understanding the fundamentals of multicomponent hydrocarbon transport in porous and dense membranes. Finally we will provide perspective on the potential for membranes to make an impact on modern refining and petrochemicals production

17a - Heterogeneous Ionization Behavior of Polyamide Thin-Film Composite Membranes for Reverse Osmosis and Nanofiltration

Jay Werber (University of Toronto)*

Polyamide thin-film composite desalination membranes are the industry standard for reverse osmosis (RO) and nanofiltration (NF). An important property is the ionization behavior of dissociable groups—particularly carboxylic acids and amines that stem from unreacted components in interfacial polymerization—which impact properties such as water uptake, water permeability, and water/ion selectivity. In their landmark work, Coronell and co-workers developed a silver-binding/Rutherford Backscattering Spectroscopy (RBS) method to demonstrate that carboxylic acids follow two dissociation constants (pKa of ~5 and ~9). However, RBS is highly inaccessible and the mechanisms of acid ionization remained unknown. Our work sought to address these issues. We first demonstrated a quenching method wherein amines and alcohols were used to neutralize residual acid chloride groups during membrane synthesis. To quantify the ionized carboxyl content, we modified the silver-binding approach to include an acidic elution step, which allowed simple quantitation using the widely available inductively coupled plasma–mass spectrometry technique. Amine content was characterized similarly, using bromide as a counter-ion. By pairing these methods with controlled membrane synthesis techniques, we demonstrated that the two pKa values for carboxyl groups stem from their location in the film: surface-bound groups ionize at the lower pKa that is similar to the bulk solution value, while interior groups ionize at the higher pKa, which we show stems from nanoconfinement and the resulting lower dielectric constant. Furthermore, ionization behavior was similar at high salinities, suggesting that interior carboxyl functionalities are neutrally charged during typical desalination processes. Our work has important ramifications for structure-property-transport relationships in these widely used materials.

17b - Counter-ion diffusion in ion-exchange membranes with varying degrees of water content

Jovan Kamcev (University of Michigan)*

Rational design of membranes with transport properties specifically tailored for a given application requires fundamental understanding of the influence of polymer structure on solute transport. Two of the most important membrane characteristics that influence ion transport in ion-exchange membranes (IEMs) are fixed charge concentration and water content. Conventional IEMs for brackish water desalination are highly swollen by water (water mass fractions between 0.4-0.5), while relatively new commercial IEMs for treating concentrated salt solutions (e.g., brines) sorb much less water (water mass fractions closer to 0.2). Ion transport in the latter class of IEMs is poorly understood on a fundamental level, which hinders our ability to rationally design high-performance membranes for such emerging applications. In this study, we systematically explored the effect of membrane water content on counter-ion diffusion in homogeneous IEMs. Cross-linked IEMs with water content that varied over a broad range were synthesized via free radical copolymerization. To decouple the effects of water content from those of fixed charge group concentration, the IEM fixed charge concentrations were held constant. Counter-ion diffusion coefficients were obtained from membrane ionic conductivity values measured via electrochemical impedance spectroscopy (EIS). The EIS measurements were performed with water-equilibrated IEMs to eliminate potential complicating effects of co-ions (i.e., only counter-ions were present within the membranes during EIS measurements). To gain deeper insights into the mechanism for counter-ion diffusion in the IEMs, activation energies for diffusion were obtained by performing the ionic conductivity measurements as a function of temperature. The role of various factors (e.g., polymer tortuosity, ion hydration, ion pairing, etc.) that potentially influence counter-ion diffusion in IEMs at various degrees of water content is the main topic of this presentation.

17c - Bottom-up synthesis of films hosting atom-thick molecular-sieving apertures

Cédric Van Goethem (École polytechnique fédérale de Lausanne)* ; Luis Francisco Villalobos (Ecole polytechnique fédérale de Lausanne); Kumar Varoon Agrawal (Ecole polytechnique fédérale de Lausanne)

Membrane-based gas separation is expected to provide an energy-efficient gas separation technique provided that membranes can be developed that combine selective permeation with large permeance. The ultimate limit to this being two-dimensional (2D), atom thick selective layers hosting a high-density of molecular sieving apertures.[1] Graphene is a highly promising 2D membrane material, but existing techniques for its synthesis with high pore density are tedious because they rely on post-synthetic creation of pores in otherwise defect-free pristine graphene.

In this work, we address this through the development of graphene crystallization conditions that promote the formation of nm and sub-nm-sized intrinsic defects.[2] Graphene films are formed through the precipitation of carbon dissolved in a Ni matrix with the carbon dissolution regulated by the controlled pyrolysis of precursors (polymers and/or sugar). The resulting porous nanocrystalline graphene (PNG) films were found to be made up of few-layer graphene, except near the grain edge where the grains taper down to single-layer and eventually terminate into vacancy defects at a node where multiple grains meet. The PNG-nanostructure and processes involved in PNG-formation were studied using TEM, NMR and TGA. The high pore density (1012 cm-2, comparable with state of post-synthetic etching strategies[1]) resulted in attractive gas separation performance with H2 permeances up to 38000 GPU and attractive molecular sieving from CH4 and N2 (at 150 °C, H2/CH4 = 17.7 and H2/N2 = 16.0). Post-synthetic masking of the PNG-pores with a CO2-phillic PEGDE-swollen PEI-layer produced membranes with favorable CO2/N2 separation (CO2 permeance of 5300 GPU and CO2/N2 selectivity of 31). Overall, the reported method improves the scale-up potential of graphene membranes by cutting down the processing steps.

[1] J. Zhao, et al., Science advances, 2019, 5, eaav1851.

[2] L.F. Villalobos and C. Van Goethem, et al., PNAS, 2021, accepted.

17d - Entrapped Nanobubbles as Ultra-selective and Oxidation-resistant Membranes for Desalination and Water Reuse

Duong T. Nguyen (University of Colorado Boulder)* ; Sangsuk Lee (University of Colorado Boulder); Kian P Lopez (University of Colorado Boulder); Mark Hernandez ("University of Colorado, Boulder"); Jongho Lee (University of British Columbia); Anthony Strau

Current reverse osmosis membranes poorly remove harmful water contaminants and are vulnerable to degradation from oxidants used in water treatment, such as chlorine and ozone. Here, we show that nanoporous hydrophobic membranes with an entrapped nanobubble layer inside their pores can facilitate pressure-driven desalination with high selectivity and oxidation resistance. Fabricated membranes with 20 nm pores demonstrate near-complete rejection (greater than 99.8%) of sodium chloride salts at operating pressures up to 34.5 bar as well as 99.1%, 98.1%, and 95.8% rejection of boron, urea, and N-nitrosodimethylamine, respectively. We show that desalination performance of the membranes is unaffected after exposure to chlorine (1000 ppm in 36 hrs) and ozone (25 ppm in 1 hr). We also find that reducing thickness allows the membranes to circumvent the permeability-selectivity trade-off that constrains conventional polymeric reverse osmosis membranes. Finally, we demonstrate the scalability of the process by desalinating water using large-area polytetrafluorethylene membranes and show biofouling prevention from Pseudomonas aeruginosa bacteria using chlorine treatment.

17e - Advancing membrane chromatography processes for the purification of therapeutic viruses

Karina Kawka (McMaster University)* ; A. Noelle Wilton (McMaster University); Maria Fe C. Medina (McMaster University); Brian Lichty (McMaster University); Raja Ghosh (McMaster University); David Latulippe (McMaster University)

Viruses are gaining significant interest as a new class of biotherapeutics being used as vectors in gene therapy and vaccines and as cancer-killing agents in immunotherapies. While these new therapies show great promise, the large-scale purification of viruses is hampered by poor overall yields (typically less than 20%); thus there is a need for novel downstream purification processes. The relatively large pore size of membrane adsorbers makes them promising materials for the removal of host cell impurities. Similar to other biotherapeutics, the downstream processes developed for virus purification involve many interconnected unit operations for which a variety of process materials are commercially available. Thus, it remains a challenge to select the optimal conditions for each process step. To address this challenge, we used an integrated approach to evaluate the effect of the DNA digestion step on the performance of Sartobind Q anion-exchange (AEX) membranes for adenovirus purification. After a series of optimizations, an 80-fold improvement in the removal of DNA and a 10-fold scale-up in the amount of processed virus were achieved. Next, the performance of two popular AEX membranes (Sartobind Q and Mustang Q) was assessed by making laterally-fed membrane chromatography (LFMC) devices containing 1 mL of each membrane. Surprisingly, this sort of direct comparison has never been reported in the literature. At a given set of elution conditions, the Sartobind Q membrane had a higher separation resolution, with 48% less DNA impurity. The collection of experimental results obtained to date are currently being integrated into an economic analysis study of virus manufacturing using the BioSolve Process software. It is anticipated that the results from this analysis will address the critically outstanding questions related to virus purification and thus enable the development of the most economical way to produce therapeutic viruses at various scales of manufacturing.

17f - Membrane Protein-Based Biomimetic Membranes for Water Treatment

Yu-Ming Tu (The University of Texas at Austin)* ; Hyeonji Oh (University of Texas at Austin); Benny Freeman (The University of Texas at Austin); Manish Kumar (University of Texas at Austin)

Biomimetic and bioinspired membranes (BBMs) containing MP channels and MP-mimic functionalized materials have emerged as innovative platforms to develop membrane-based separations and purification with specific solute selectivity [1]. BBMs, utilizing biological elements or mimicking concepts from biological mechanisms, can take advantage of processes evolved by nature for enhancing transport efficiency and specificity. The MP channels incorporated have well-organized and uniform inner pore diameters, generating a complete uniformity of pore size in BBMs, in direct contrast to the widely pore size distribution in current commercial membranes [2]. MP-based membranes demonstrate specific small solute selectivity and exhibit greater water permeability than commercial nanofiltration membranes with similar separation ranges [3]. This research also explores affordable and sustainable solutions for creating tight MP-based membranes with fewer defects. We developed defect sealing strategies to fabricate BBMs with a single deposition step of MP-block copolymer (BCP) crystals combined with defect-filling materials, UV cross-linkable phospholipids-23:2 Diyne PC. The defect-sealed MP membranes demonstrate >7-log MS2 viral particle removal, exceeding 4-log removal suggested by US Environmental Protection Agency (EPA) for water treatment. We will apply this fabrication process to create BBMs reconstituted with aquaporin channels for water desalination.

[1] Song, W., Tu, Y. M., Oh, H., Samineni, L., & Kumar, M. "Hierarchical optimization of high-performance biomimetic and bioinspired membranes." Langmuir (2018).

[2] Tu, Y. M., Samineni, L., Ren, T., Schantz, A. B., Song, W., Sharma, S., & Kumar, M. "Prospective Applications of Nanometer-scale Pore Size Biomimetic and Bioinspired Membranes." Journal of Membrane Science (2020).

[3] Tu, Y. M.†, Song, W.†, Ren, T.†, et al. "Rapid fabrication of precise high-throughput filters from membrane protein nanosheets." Nature Materials (2020).

18a - Decoupling ionic conduction and cross-over in membrane separators for non-aqueous redox flow batteries

Geoffrey M Geise (University of Virginia)*

Providing sustainable supplies of clean energy is a critical global challenge for the future. Renewable energy generation technologies are likely an important part of the solution, but often these technologies generate electricity intermittently. Key to addressing this challenge is the ability to store and deliver grid-scale amounts of energy, and redox flow batteries are a promising option for doing just that. One strategy for enhancing redox flow battery energy density is to use a non-aqueous targeting redox flow battery approach. In this battery, the membrane separator, exposed to organic solvent, must permit ionic conduction while simultaneously restricting passage (or cross-over) of organic redox active molecules (or shuttles).

A functionalized poly(phenylene oxide) cation exchange material has shown promise as a membrane separator for non-aqueous redox flow battery applications [1]. This material permits conduction of lithium ions but restricts passage of representative redox active molecules (e.g., metallocenes, aromatic organics, and aliphatic organics). Analysis of thermodynamic interactions between the polymer membrane and the probe molecules provide a possible explanation for an observed decoupling of the ionic conductivity and organic molecule cross-over properties of the membrane as the degree of functionalization is manipulated. This presentation will discuss lithium ion conduction and selectivity structure-property-processing relationships in a poly(phenylene oxide)-based charged polymer for emerging non-aqueous redox flow battery applications. The presentation also will discuss the importance of interactions that appear to be significant in non-aqueous systems compared to aqueous applications.

[1] McCormack, Luo, Geise, Koenig. Journal of Power Sources 2020, 460, 228107.

18b - Independent tuning of anion exchange membrane conductivity and permselectivity via non-covalent crosslinking

Ryan S Kingsbury (University of North Carolina at Chapel Hill)* ; Jingbo Wang (University of California Los Angeles); Maruti Hegde (University of North Carolina at Chapel Hill); Ahmet Kusoglu (Lawrence Berkeley National Lab); Wei You (University of North C

Ion exchange membranes (IEMs) are a key component of many electrochemical processes for clean water and energy production, including batteries, fuel cells, electrolyzers, water purification by electrodialysis, and salinity-gradient energy technologies such as reverse electrodialysis. For optimum process performance, it is highly desirable that these membranes possess both high conductivity and high permselectivity toward the ion(s) of interest.

Most conventional polymer IEMs are covalently crosslinked, which results in a challenging tradeoff relationship in which high permselectivity and high conductivity are difficult to achieve simultaneously. In an attempt to overcome this limitation, we synthesized a series of anion exchange membranes containing non-covalent crosslinks formed by a hydrogen bond donor (methacrylic acid) and a hydrogen bond acceptor (dimethylacrylamide). We found that the hydrogen bond donor and acceptor act synergistically to improve both membrane permselectivity and conductivity relative to a control membrane without non-covalent crosslinks. Furthermore, we show that the hydrogen bond donor and acceptor loading can be used to tune permselectivity and conductivity relatively independently of one another, escaping the tradeoff observed in conventional membranes containing only covalent crosslinks.

Our observations cannot be rationalized via the conventional understanding of the conductivity-selectivity tradeoff, which is based on membrane swelling and its effect on Donnan exclusion. As such, our results suggest that our understanding of the physicochemical factors governing ion sorption and transport in charged membranes is incomplete. We investigate the states of water and the microstructure inside the synthesized IEMs to achieve a more complete understanding of their behavior, and comment on possible mechanisms involved.

18c - Catalytic reactions power the self-pumping membrane

Yuhang Fang (Purdue University)* ; Jeffrey Moran (George Mason University); David Warsinger (Purdue)

Catalytic micromotors, such as Pt-Au metallic rods, use chemical energy in catalytic reactions to propel themselves in liquid. By fixing such motors onto the surface of membrane, here we show with simulation results, Pt-Au coated membrane could pump fluid through the pores via catalytic reactions. We demonstrate that this self-pumping flow is an electroosmotic flow driven by the self-generated electric field due to catalytic reactions. By creating concentration gradient of H+, catalytic reactions self-generate electric field, which drives the electric double layers carry the bulk fluid through the pores. With small pore radius, such as 0.5um, the electrical double layers are overlapped, which limits the self-pumping flow; with large pore radius, such as 6um, a self-pumping flow of 23um/s could be achieved, which is 20 times higher than the reported experimental results. This work highlights the potential of using catalytic reactions to pump liquid through membranes without external powers. Meanwhile, we enhance the understanding of self-pumping flow and provide guidance on the designing next generation of self-pumping devices.

18d - Revisiting Water and Ion Transport in Nafion

Rahul Sujanani (The University of Texas at Austin)* ; Benny Freeman (The University of Texas at Austin)

Nafion is an ion exchange polymer of significant interest as a fuel cell membrane due to rapid transport of protons, often ascribed to its phase-separated morphology. The transport of water and ions in Nafion has been explored in considerable detail over the past several decades and draws continued interest. However, the Nafion literature has evolved somewhat independently from other fields studying charged polymers (e.g., polyelectrolytes and water purification). For example, there are conflicting reports about the fundamental physics of water transport in Nafion, with some authors postulating a pore flow mechanism and others applying the solution/diffusion model. Moreover, many studies of ion transport in Nafion focus on proton transport, while most studies in charged polymers in the membrane community focus on ions other than protons, given the importance of such ions to desalination. This presentation will detail experimental measurements and theoretical modeling of water and ion transport in Nafion, connecting teachings from several communities interested in the transport of neutral and charged solutes through polymers. Topics addressed include the influence of ion association on salt sorption, the importance of counter-ion condensation on ion transport, the transport mechanism of uncharged solutes (i.e., water), and the utility of existing models (and gaps therein) to describe thermodynamics and transport phenomena in Nafion and related ionomers.

18e - Screening of electrostatic interactions between carboxylates and ion exchange membranes by co-transporting alcohols

Luca Kim (Auburn University)* ; Yihung Lin (Auburn University); Bryan S Beckingham (Auburn University)

Understanding multi-component transport behavior through hydrated dense membranes is of interest for numerous applications. For the particular case of photoelectrochemical CO2 reduction cells (PEC-CRC), it is important to understand the multi-component transport behavior of CO2 electrochemical reduction products including mobile carboxylates (formate and acetate) and alcohols (methanol and ethanol) in the ion exchange membranes (i.e. anion exchange membranes, AEM, and cation exchange membranes, CEM) as one role of the membrane in these devices is to minimize the permeation of these CO2 reduction products to the anolyte as they often oxidize back to CO2. Both AEMs and CEMs have their advantages and drawbacks for such devices. For AEMs, they act to facilitate the transport of common electrolytes, such as bicarbonates, but they also act to facilitate the transport of carboxylates. For CEMs, they act to minimize the transport of carboxylates, but the transport of carboxylates often increases in co-permeation with alcohols. Therefore, the design of new AEMs and CEMs is necessary. Here, we investigate the transport behavior of carboxylates and alcohols in two types of AEMs (Selemion®AMVN and a crosslinked PEGDA-APTA), a charge-neutral crosslinked PEGDA, and two types of CEMs (Nafion®117 and a crosslinked PEGDA-AMPS). In both AEMs and CEMs, we observe changes in transport behavior which we ascribe to changes in the electrostatic interactions. In AEMs, we observe a decrease in diffusivities to carboxylates in co-diffusion with alcohols, where we attribute this behavior to charge screening by co-diffusing alcohol that reduces the electrostatic attraction between bound quaternary ammoniums and mobile carboxylates. In CEMs, we observe an increase in diffusivities to carboxylates in co-diffusion with alcohols, where we attribute this behavior to charge screening by co-diffusing alcohol that reduces the electrostatic repulsion between bound sulfonates and mobile carboxylates.

18f - Elucidating Counterion Mobility in Ion-Exchange Membranes: Spatial Effect and Valency-Dependent Electrostatic Interaction

Hanqing Fan (Columbia University)* ; Yuxuan Huang (Columbia University); Ian Billinge (Columbia University); Ngai Yin Yip (Columbia University)

Ion-exchange membranes (IEMs) are highly-charged polymeric thin-films widely applied in energy and water technologies, such as electrodialysis. Traditional IEMs processes utilize the charged membranes to allow the permeation of oppositely charged counterions and block like-charged co-ions. Recently, separation beyond the charge polarity, i.e., ion-specific separation, has attracted increasing research interest. The counterion-specific selectivity can unlock new applications, e.g., selective extraction of valuable elements from seawater and recovery of nutrients in wastewater treatment. However, gaps in the current understanding of the chemistry-structure-property relationships in membrane ion mobility still exist, frustrating efforts on informed development of ion-specific selectivity. In this study, we propose a theoretical framework to investigate the migration mobility of different counterions in IEMs. The model establishes a quantitative structure-property relationship, based on two governing effects: the spatial effect of swollen water and the electrostatic interaction between mobile ions and fixed charges. The counterion diffusivities of commercial IEMs in different electrolytes were experimentally characterized and show good agreement (high regression R2) with the proposed model. The impact of counterion-dependent water swelling is assessed for ion-specific separation. The frictional effect of electrostatic interaction is quantitatively linked to the structural and material properties of fixed charged density and effective dielectric constant. The analysis also reveals that anion exchange membranes have a weaker electrostatic effect compared to cation exchange membranes, due to differences in the molecular features intrinsic to the membranes. This study offers new insights on the underlying mechanisms governing ion-specific selectivity and will inform the rational application of highly-charged polymeric thin-films for high-value targeted ion separations.

19a - Ultra-permeable wafer-scale SWCNT membranes

Melinda L Jue (Lawrence Livermore National Laboratory)* ; Steven Buchsbaum (Lawrence Livermore National Laboratory); Chiatai Chen (Cornell University); Sei Jin Park (Lawrence Livermore National Laboratory); Eric Meshot (Lawrence Livermore National Laborato

Single-walled carbon nanotubes (SWCNT) promise to enable major advancements in membrane applications such as water purification, low-cost recovery of high-value components, and advanced protective garments through their enhanced fluid transport. Practical applications of these membranes have been hampered by challenges in fabricating large-area materials with a high density of open, small-diameter, SWCNT pores. High tube density is required to outperform the flow rates seen in commercial membranes and small tube diameters enhance both size- and charge-based selectivity. Here, we optimized the growth of vertically aligned SWCNTs to maximize the density (up to 2x1012 tubes/cm2), minimized the average diameter (to less than 2 nm), and simultaneously scaled up the growth area (up to 100 mm in diameter). Ultra-permeable large-area SWCNT membranes were fabricated by filling the interstitial space with a chemically resistant polymer followed by removal of the SWCNT caps with reactive ion etching. We demonstrate wafer-scale membranes with water permeances up to 250 LMH/bar, which greatly surpass commercial loose nanofiltration/tight ultrafiltration membranes. Additionally, these SWCNT membranes display high rejection of dyes (e.g., Rose Bengal) while simultaneously permitting complete passage of salts such as NaCl and Na2SO4. Contrary to conventional membranes for water treatment, aggressive cleaning methods did not detrimentally affect the SWCNT membrane performance. Together with their ultrahigh permeance, the demonstrated chemical resistance of these membranes offers opportunities for energy-efficient nanofiltration/ultrafiltration processes in chemically demanding environments.

19b - MF and UF Coated Membranes for Selective Separation of Organic Anions- PFAS and Trivalent Cations

Francisco Leniz (university of kentucky)* ; Dibakar Bhattacharyya (U. Kentucky); Ronald Vogler (University of Kentucky); Phillip Sandman (University of Kentucky); Natalie Harris (University of Kentucky)

Contamination of water sources due to mankind's processes and the release of degraded materials into the environment have become an increasing concern over the past decades. From the extraction of the organic anions per- and polyfluoroalkyl substances (PFAS) to the recovery of the valuable inorganic cations lanthanides, nanofiltration (NF) membranes are a suitable candidate for selective separation of these components. Therefore, to optimize this process a critical assessment of the complexity of the separation of these compounds from the display of their properties in the aqueous solution to their partitioning into the polymeric layer is needed.

PFAS and model organic anions such as their hydrogenated counterpart among others were systematically studied. Bringing up molecular properties and their display in aqueous solution allowed a comprehensive understanding of the separation of these compounds. Meanwhile some PFAS performed more stable high retentions, other organic anions such as oxalate can transition from approximately 100% rejection down to 0% rejection by increasing the hydrogen concentration of the solution by 4 orders of magnitude. Ultimately, a single sorption/exclusion platform for PFAS separation was created enabling the potential of truly low concentration filtered water.

Valuable inorganic cations such as lanthanum and neodymium were also studied. The potential of high selective separation at low permeate flux using positively charged NF membranes was elucidated using the Donnan steric pore model with dielectric exclusion (DSPM-DE). Synthesized positively charged NF membranes using polyallylamine performed high and stable lanthanides retentions, with values around 99.3% in mixtures with high ionic strength (100 mM, equivalent to ~6000 ppm), 99.3% rejection at 85% water recovery (and high Na^+/La^{3+} selectivity, with 0% Na^+ rejection starting at 65% recovery), and both constant lanthanum rejection and permeate flux at even pH 2.7.

19c - Decoupling entrance and inner resistances in CNT channel

Melinda L Jue (Lawrence Livermore National Laboratory); Steven Buchsbaum (Lawrence Livermore National Laboratory); Sei Jin Park (Lawrence Livermore National Laboratory); Kathleen Moyer (Lawrence Livermore National Laboratory); Francesco Fornasiero (Lawren

Due to their high transport rates, a great deal of attention has been recently given to 2D materials and slippery 1-D nanochannels as promising building blocks for next generation membranes. While in 2-D materials high flux is expected because of the classical inverse scaling of the flow rate with the pore thickness, in smooth channels such as carbon nanotubes (CNT), orders of magnitude rate enhancements with respect to classical theories are attribute to a vanishing friction at the pore wall. Irrespectively of the high flux origin, in both atomically thin and thicker but slippery nanopores, the flow rates are largely dictated by the entrance/exit hydrodynamic resistance.

For CNT channels, experimental quantification of the magnitude of end and inner resistances is still lacking despite its importance for both practical applications and fundamental understanding. This has led to inaccuracy and disagreement in the calculation of slip lengths and flow rate enhancements from experimentally measured permeation rates, since often entrance/exit resistances are neglected altogether, or an arbitrary magnitude is assumed. Here, we quantified these resistances for both gases and liquids in CNT channels by fabricating membranes with controlled CNT length and known number of open pores. We found that the end resistance dominates the total resistance. For liquid water, measured viscous energy dissipation at the nanotube ends is quantitatively described by Sampson equation. For 2.4 nm wide single-walled CNTs, measured slip lengths approach several microns. A prevailing contribution of the end resistance was also found in pressure-driven gas transport, and recorded flow rate enhancements with respect to Knudsen theory appear to be independent of the gas type. These findings further advance the community understanding of the peculiar and often unusual CNT fluidic properties and may help reconciliating "conflicting" literature reports on the subject.

19d - Atomic layer deposition onto and within polymers for controlling interfaces and nano-structuring UF membranes

Tamar Itzhak (Technion- Israel Institute of Technology); Assaf Simon (Technion); Zhenzhen Zhang (Helmholtz-Zentrum Geesthacht); Naama Segev-Mark (Technion); Clarissa Abetz (Helmholtz-Zentrum Geesthacht); Volker Abetz (Helmholtz-Zentrum Geesthacht); Guy Z

Atomic layer deposition (ALD) is a powerful technique for thin layer deposition of inorganic materials on tortures, high aspect ratio, substrates. By using ALD, a wide variety of inorganic compositions can be deposited on organic and inorganic substrates with sub-nm accuracy. Thus, ALD has emerged in recent years as a new methodology for controlling the membranes' interfaces.

In our research, we investigate ALD onto and within polymers. We show, using high-resolution electron microscopy, that controlling the diffusion time of ALD precursors dictates the metal oxide deposition through the depth of the polymeric, phase inverted, ultrafiltration membranes. With just several cycles of Al2O3 ALD, we were able to modify the polymer membranes' interface, enhancing the membrane hydrophilicity, and oil anti-fouling behavior.

We further explore ALD within polymers, also known as sequential infiltration synthesis (SIS). In SIS, long diffusion times and precursor-polymer reactivity enables metal oxide growth within the polymers' volume, creating organic-inorganic composite materials. We performed SIS on self-assembled block copolymer (BCP) layers, where the polymer block chemistry yields selective SIS metal oxide growth in only one block. Following the growth, we removed the BCP template to obtain isoporous metal oxide membranes with enhanced pollutant selectivity.

Finally, we combine both approaches to tune the pore size and surface properties of anisotropic BCP membranes (SNIPS) where we selectively grow metal oxides in the polymer block surrounding the pore with SIS and further perform ALD cycles to tune pore size from ultrafiltration to nanofiltration. The metal oxide pore interface allows us to control the filtration selectivity as well as be the basis for new functional moieties addition.

19e - Ion transport in sub-1-nm carbon nanotube porins

Aleksandr Noy (Lawrence Livermore National Laboratory)*

Controlling water and ion transport on a molecular scale is important for membrane applications ranging from industrial water treatment, to precision separations, to energy production. Living systems are adept at moving ions and small molecules across biological membranes using protein pores that rely on nanoscale confinement to achieve efficient and exquisitely selective transport. I will show that carbon nanotube porins—pore channels formed by ultra-short carbon nanotubes assembled in a lipid membrane—can exploit similar physical principles to transport ions and achieve differential ion selectivity. I will discuss the role of molecular confinement and ion interactions with the pore walls and show how these phenomena can enhance water and ion transport efficiency and influence the mechanisms of ion selectivity and electroosmotic coupling in these pores. Overall, carbon nanotube porins represent simple and versatile biomimetic membrane pores that are ideal for fundamental nanofluidics studies and for building the next generation of membrane separation technologies.

19f - Improving the Long-Term Performance of Living Filtration Membranes

Emily Rutledge (Montana Technological University)*

Living filtration membranes (LFMs) have shown to be effective ultrafiltration membranes, using SCOBY (Symbiotic Culture of Organic Bacteria and Yeast), a bacterial cellulose. They are also presented as a more affordable and sustainable alternative to synthetic filtration membranes. A great challenge is the relatively short lifespan. In this study, the membranes undergo chemical treatments for extending the lifespan of operation through crosslinking. Membranes treated with 5% and 10% citric acid solutions were compared with pristine membranes. Crosslinking membranes with citric acid solutions and the addition of a sodium hydroxide base treatment, will form ester bonds between cellulose fibers in the membrane structure. These bonds should strengthen the membrane structure and allow the membrane to maintain ultrafiltration characteristics over a longer period of time. Crosslinking was successful after the Fourier-Transform Spectroscopy (FTIR) analysis and thermogravimetric analysis (TGA) methods. The presence of the carbonyl and ester functional groups were represented by peaks on FTIR charts at wavelengths around 1700 and 1590 cm-1. Using TGA, the presence of citric acid was shown as a decomposition peak between 100° and 200°. The surface charge will be measured using a SURPASS3 streaming potential analyzer, and will then be compared to the pristine and MCE membranes that both have relatively negative surface charges. Based on contact angle goniometry, treated membranes maintained the hydrophilicity that has been demonstrated in pristine living filtration membranes as well as mixed cellulose ester (MCE) membranes. The membrane samples that underwent the crosslinking process were typically 1-1.5 mm thicker than a commercial ultrafiltration membrane, so the permeability rate is likely to be slower. Continued testing will allow for more concrete comparisons of the lifespan of treated living filtration membranes compared to pristine and mixed cellulose ester membranes.

20a - Performance of tangential flow filtration using reverse asymmetric membrane for CHO cell harvesting

Da Zhang (University of Arkansas); Xianghong Qian (University of Arkansas); Parag Patel (Asahi Kasei); Daniel Strauss (Asahi Kasei); April Wheeler (Asahi Kasei); Ranil Wickramasinghe (University of Arkansas)*

Tangential flow filtration is advantageous for bioreactor clarification as the permeate stream could be introduced directly to the subsequent product capture step. However, membrane fouling coupled with high product rejection has limited its use. Here, the performance of a reverse asymmetric hollow fiber membrane where the more open pore structure faces the feed stream and the barrier layer faces the permeate stream has been investigated. The open surface contains pores up to 40 µm in diameter while the tighter barrier layer has an average pore size of 0.4 µm. Filtration of Chinese hamster ovary cell feed streams has been investigated under conditions that could be expected in fed batch operations. The performance of the reverse asymmetric membrane is compared to that of symmetric hollow fiber membranes with nominal pore sizes of 0.2 and 0.65 µm. Laser scanning confocal microscopy was used to observe the locations of particle entrapment. The throughput of the reverse asymmetric membrane is significantly greater than the symmetric membranes. The membrane stabilizes an internal high permeability cake that acts like a depth filter. This stabilized cake can remove particulate matter that would foul the barrier layer if it faced the feed stream. An empirical model has been developed to describe the variation of flux and transmembrane pressure drop during filtration using reverse asymmetric membranes. We have also attempted to identify specific species that foul the membrane. Our results suggest that using a reverse asymmetric membrane could avoid severe flux decline associated with fouling of the barrier layer during bioreactor clarification.

20b - Scalable synthesis of nanoporous atomically thin graphene membranes for dialysis and molecular separations via facile isopropanol-assisted hot lamination

Piran Kidambi (Vanderbilt University)*

Scalable graphene synthesis and facile large-area membrane fabrication are imperative to advance nanoporous atomically thin membranes (NATMs) for molecular separations. Although chemical vapor deposition (CVD) allows for roll-to-roll high-quality monolayer graphene synthesis, facile transfer with atomically clean interfaces to porous supports for large-area NATM fabrication remains extremely challenging. Sacrificial polymer scaffolds commonly used for graphene transfer typically leave polymer residues detrimental to membrane performance and transfers without polymer scaffolds suffer from low yield resulting in high non-selective leakage through NATMs. Here, we systematically study the factors influencing graphene NATM fabrication and report on a novel roll-to-roll manufacturing compatible isopropanol-assisted hot lamination (IHL) process that enables scalable, facile and clean transfer of CVD graphene on to polycarbonate track etched (PCTE) supports with coverage ≥99.2%, while preserving support integrity/porosity. We demonstrate fully functional centimeter-scale graphene NATMs that show record high permeances (~2–3 orders of magnitude higher) and better selectivity than commercially available state-of-the-art polymeric dialysis membranes, specifically in the 0–1000 Da range. Our work highlights a scalable approach to fabricate graphene NATMs for practical applications and is fully compatible with roll-to-roll manufacturing processes.

References:

1. Cheng P. et al. Nanoscale 2021

2. Kidambi et al. Adv. Mat. 2018

3. Kidambi et al. Adv. Mat. 2017

20c - Rapid size and affinity based detection of intact viral particles using functionalized microslit silicon membranes

Michael Klaczko (University of Rochester)*; Kilean Lucas (University of Rochester); Alec Salminen (University of Rochester); Brian Ward (University of Rochester Medical Center); Jonathan Flax (University of Rochester); James McGrath (U. Rochester)

The COVID-19 pandemic clearly demonstrated the public health benefits of reliable and accessible diagnostic tests for viral infections. Despite the rapid development of gold-standard polymerase chain reaction (PCR) for the SARS-CoV-2 virus, global demand created logistical challenges that delayed access to testing for months. Additionally, the extreme sensitivity of PCR had a costly downside as the tests could not differentiate between patients with active infection and those who were no longer infectious but still shedding viral fragments. To address these issues, we propose a membrane-based sensor that requires no external power and detects only intact virus. The sensor employs affinity and sized based capture onto an ultrathin microslit membrane. The presence of intact virus, but not viral debris, fouls the membrane to trigger a hydraulic switch that can be visualized after the injection of a 40 µL sample with a pipette. The device, which we call the µSIM-DX (microfluidic device featuring a silicon membrane for diagnostics), features a biotin-coated microslit membrane with pores ~2-3X larger than the intact virus. Streptavidin-conjugated antibody recognizing viral surface proteins is incubated with the sample for ~1 hour before injection. If present, intact virus is captured and triggers the re-routing of the injected sample providing a visual indicator of a positive test. Proof-of-principle tests have been done using vaccinia virus. After optimizing slit pore sizes, the fouling-based sensor exhibits 100% specificity and 96% sensitivity for vaccinia virus (n=34). Moreover, the dynamic range of the sensor extends from 10^5-10^9 copies/mL which encompasses the entire range of the mean viral loads for SARS-CoV-2 in COVID-19 (10^5.6-10^7 copies/mL). Future work will test the ability of our sensor to perform in biological fluids and with SARS-CoV-2, demonstrating its potential as a PCR-free alternative for containment efforts in the spread of infectious disease.

20d - Factors Affecting Robustness of Anion Exchange Chromatography: Selective Retention of Minute Virus of Mice Using Membrane Media

Wenbo Xu (University of Arkansas)*; Shu-Ting Chen (University of Arkansas); Kang Cai (AstraZeneca); Gisela Ferreira (AstraZeneca); Ranil Wickramasinghe (University of Arkansas); Xianghong Qian (University of Arkansas)

Mobile and stationary phase factors were investigated in order to identify conditions for effective capture of minute virus of mice (MVM), a potential adventitious contaminant in biomanufacturing, using anion exchange membrane chromatography (AEX). The initial study was conducted for Membrane A for a range of feed conditions using bovine serum albumin (BSA) as a model protein mimicking acidic host-cell proteins (HCPs). The effects of pH (~6-8), salt concentration (~0-150 mM NaCl) and level of BSA (~0-10 g/L) were systematically investigated. It was found that higher BSA concentration has the most negative impact on MVM binding followed by the increased conductivity of the feed solution. The effect of pH on MVM binding is also detected but has a less impact compared to other two factors in the range of feed conditions investigated. In addition to Membrane A, three other AEX membranes (Membrane B, C and D) were investigated for MVM binding at a selected feed condition. Based on properties of the membranes investigated, it was found that ligand charge density has the most significant impact on MVM binding performance of AEX membranes from stationary phase perspective.

20e - Catalytic DNA-membrane reactor

Beñat Olave (Polymat, University of the Basque Country); Iván Rivilla (DIPC); Fernando Cossío (University of the Basque Country); Thomas Schäfer (Polymat, University of the Basque Country)*

One way to approximate the ingenuity of biological membranes while maintaining the robustness of artificial membranes is the use of hybrid materials. We have studied modular systems comprising functional-DNA and mesoporous anodized aluminum oxide (AAO). Functional nucleic acids are highly versatile molecular building blocks: thanks to their specific hybridization they enable systematically creating nanoscale structures; recognize a wide variety of molecules; or even serve as biocatalysts [1]. Recently, DNA was shown to be capable of catalyzing a 1,3-dipolar reaction which does otherwise not exist in living systems [2]. However, commonly the applications of functional nucleic acids are limited to aqueous environments, because molecular organic solvents widely annihilate their function or their thermal or conformational stability.

In this study we made use of the intrinsic property of DNA-modified membranes of creating a reactive interface between compartments of different chemical composition, similarly to what biological membranes do in a far more ingenious manner under physiological conditions. For this purpose, AAO membranes were functionalized with DNA whose chirality can be transferred directly to a metal-catalyzed reaction, and in particular a copper(II)-catalyzed Diels–Alder reaction. In this way, we created a prototype of what might be denominated a catalytic DNA-membrane reactor.

We studied the immobilization efficiency and functional stability of DNA under prolonged operation of the reactor and compared the results with homogenous DNA-catalysis. The results suggest that we can use versatile functional molecules such as nucleic acids beyond physiological conditions and in a DNA-membrane reactor which offers mechanical robustness and the specific enzymatic activity of DNA.

[1] G. Roelfes and B.L. Feringa, Angewandte Chemie, 2005, 3230-3232

[2] I. Rivilla et al., Chem Sci., 2017,7038-7046

20f - Membrane emulsification for the preparation of uniform functionalized droplets with catalytic properties

Emma Piacentini (National Research Council of Italy, Insitute on membrane Technology, CNR-ITM); Rosalinda Mazzei (National Research Council of Italy, Insitute on membrane Technology, CNR-ITM); Lidietta GIORNO (National Research Council of Italy - Institut

Membrane emulsification has unique properties in terms of precise manufacturing of emulsion droplets in mild operative conditions suitable to preserve the stability of bioactive labile components. The uniformity of the droplet size is the reflection of the uniformity of pore size distribution. The formation of uniform microdroplets functionalized with interfacial catalyst promotes optimal macromolecular conformation as well as mass transfer of reagents between the two immiscible phases. The catalytic performance of interfacial biocatalyst distributed at the interface of droplets or solid capsule membranes obtained by non-solvent induced phase separation will be presented.

SPG membranes have been used to prepare emulsions. The system has been tested using lipase from Candida rugosa that served also as emulsifier. The kinetic resolution of naproxen methyl ester was used as a reaction model. Lipase showed maximum enantioselectivity (100 %) and conversion in the hydrolysis of S-naproxen methyl ester when membrane emulsification technique was used for biocatalytic microdroplets production.

The controlled fabrication of microdroplets permitted to estimate the hydrodynamic radius of the enzyme at the interface basing on enzyme mass and emulsified dispersed phase volume. Basing on research contributed by the authors, the perspective of membrane emulsification in various industrial sectors will be outlined.

21a - Centrifugal Reverse Osmosis – A Novel Membrane Model Configuration for Desalination Near Local Thermodynamic Equilibrium

William B Krantz (University of Colorado at Boulder)* ; Tzyy Haur Chong (Singapore Membrane Technology Centre)

Desalination via RO can increase the world's freshwater resources at a cost typically 2.5 times that of surface-water. This can be reduced significantly by operating RO at the thermodynamic restriction defined by local thermodynamic equilibrium between saltwater and pure water on opposite sides of a membrane. However, operation at the thermodynamic restriction requires an infinite number of stages, which seemingly is impossible. The closed-circuit RO (CCRO) process approaches operation at the thermodynamic restriction by continually recirculating the retentate in an RO module while gradually increasing the pressure until the desired recovery is achieved. CCRO requires continuously adding fresh feed to the recirculating retentate to maintain a flowrate that minimizes concentration polarization and scaling that incurs a heat-of-mixing penalty. CCRO is a batch process, thereby limiting it to smaller scale RO.

An alternative to CCRO is the novel centrifugal reverse osmosis (CRO) process that consists of a parallel stack of circular flat sheet membranes, feed channels, and spacer channels attached to a feed tube that is rotated to create a centrifugal pressure that increases with radial distance from the axis-of-rotation. CRO is a continuous process applicable to both large- and small-scale desalination. Energy recovery is achieved by jetting the retentate from the rotating assembly onto stationary vanes. The recovered pressure energy is used to rotate the CRO module and to generate electricity for the pump used to pre-pressurize the feed to the minimum required to initiate RO. The jet-and-vane assembly is an energy-recovery device integral to the CRO module. Model predictions indicate that CRO can reduce the specific energy consumption relative to SSRO by 31% for 50% recovery at 56 bar for a typical 35 g/L seawater feed. A transmembrane pressure progressively increasing from 28 to 56 bar is achieved with a CRO module having a radius of 0.72 m rotating at 1000 rpm.

21b - Energy from Carbon Dioxide: A New Concept for Power Generation from Sweep Gas Permeation

Sarah Moussaddy (Oakland University)* ; Jonathan Maisonneuve (Oakland University)

Large amounts of chemical potential energy are available from concentration gradients of gas mixtures released from thermal power plant exhaust. For example, the energy available from cooling tower and flue gases of a typical coal-fired power plant is equivalent to 1-2 % of the plant's power capacity. Recovering energy from exhaust gases may open up new opportunities for improving power plant efficiency and sustainability.

In this presentation, we demonstrate operation of a novel concept for generating useful mechanical work via sweep gas permeation across a membrane. The process takes advantage of partial pressure gradients of carbon dioxide and other gas gradients to spontaneously drive mass transfer across a membrane towards a mechanical load such as a power turbine for useful energy conversion. Using a simple binary gas mixture of nitrogen and carbon dioxide and a custom laboratory bench equipped with a commercial polydimethylsiloxane membrane, we demonstrate power generation of up to 4.5 W/m2 from 192 g/m2/h of carbon dioxide flux observed at 2 bar applied pressure difference. The effect of reverse gas leakage across an imperfectly selective membrane is studied, and is observed to significantly reduce power density, with a maximum of 3 W/m2 from net gas flux observed in the laboratory. We overview fundamental transport dynamics and validate a finite element model which is then used to show the potential of this concept with a range of membranes, process conditions, and applications.

21c - Facilitated Transport Membranes for H2 Purification from Coal-Derived Syngas: A Techno-Economic Analysis

Yang Han (The Ohio State University)* ; Winston Ho (OSU)

A single-stage membrane process is designed for using facilitated transport membranes (FTMs) to decarbonize the coal-derived syngas from an integrated gasification combined cycle (IGCC) power plant. The necessary process model and costing method are developed to assess the technical feasibility and process economics. In order to account for the carrier saturation phenomenon associated with FTMs, a homogeneous reactive diffusion model is integrated into the process model. The techno-economic study reveals that the mitigated carrier saturation upon bulk CO2 removal can lead to appreciable increases in the CO2 permeance and CO2/H2 selectivity, which can be utilized to achieve 95% CO2 purity and 95% H2 recovery with a CO2/H2 selectivity of 50 at the complete carrier saturation. In comparison, non-reactive, solution-diffusion membranes are unsuitable for the single-stage membrane process. A CO2/H2 selectivity of 85 is required for 95% CO2 purity, while a prohibitively high CO2/H2 selectivity of 300 is needed for >99% H2 recovery. The optimal feed pressure is 30–40 bar in order to condition the syngas to a temperature (100–120°C) that is feasible for the FTM operation. In this case, the optimal CO2 permeance at the complete carrier saturation is ca. 150 GPU. FTMs with different facilitated transport characteristics can also be arranged in a hybrid membrane configuration to render a H2 recovery of 99% and a cost of electricity of $118.5/MWh, which is 12.5% lower than that of the benchmark Selexol process. In addition, a H2S/CO2 selectivity of 4 is required to reduce the H2S content below 10 ppmv in the treated syngas, which can be directly fed to the combustion turbine without additional syngas sweetening.

21d - Membrane Applications in Biogas Upgrading and Purification

David Hasse (Air Liquide)* ; Sandeep Karode (Air Liquide); Noemi Collado (Air Liquide)

Introduction

Air Liquide uses multiple membrane technologies to purify biogas. With over 15 years of upgrading experience, Air Liquide has industry leading in house solutions. This talk will discuss the matching of membrane materials to specific separations involved in upgrading.

Biogas purification and upgrading is a fast growing field for membranes. Biogas has a wide variety of sizes and sources, resulting in the need for different strategies to handle local levels of impurities and contaminants.

Air Liquide has over 100 installed systems, and owns/operates a number of biogas sites at landfills and digesters. Systems range from 50-10,000 SCFM. The main separation is removal of carbon dioxide from methane. Membrane solutions have been used to reduce VOC's, H2S, oxygens, siloxanes and nitrogen.

This presentation will detail the technology of this separation. Systems can contain 2-5 membrane stages with multiple styles of membrane materials.

22a - Study of mineral fouling in synthetic effluent nanofiltration using real-time particle imaging and electrical impedance spectroscopy

Chidiebere S Nnebuo (Ben-Gurion University of the Negev)* ; Oded Nir (Department of Desalination and Water Treatment, Zuckerberg Institute for Water Research, The Jacob Blaustein Institutes for Desert Research, Ben-Gurion University of the Negev, Beersheba

Mineral fouling is a major hindrance to high recovery effluent nanofiltration, with calcium phosphate (CaP) and calcium carbonate (CaCO3) the most prevalent mineral foulants. In this study, we used a novel combination of real-time inline microscopy, electrical impedance spectroscopy (EIS), post SEM analysis and filtration metrics (water flux and rejection) to study mineral fouling mechanisms of CaP and CaCO3 salts in a synthetic effluent nanofiltration. We used a nanofiltration (NF) polyelectrolyte multilayer (PEM) membrane, prepared by static layer-by-layer (LbL) coating of a cationic polymer - polydiallyl dimethylammonium chloride, and anionic polymer - poly styrenesulfonate (six bi-layer) on a polyethersulfone (PES) ultrafiltration (UF) membrane. Increasing permeate recovery over filtration time was simulated through additions of CaCl2 with NaHCO3 or NaH2PO4/Na2HPO4. Using the novel combination of methods, we delineated the mechanisms governing fouling development with time for both CaCO3 and CaP. For CaCO3, a transition from heterogeneous precipitation on the membrane surface (scaling) to particulate fouling due to bulk precipitation was identified. For CaP, a transition between fouling by amorphous particles to fouling by crystalline particles was identified; and this phase-change was captured in real-time images using in-line microscope. We also found that for similar precipitation potentials measured by weight, CaP fouling was more detrimental to water flux (86% decrease) compared to CaCO3 (20% decrease), due to the voluminous amorphous phase. We established inline microscopy as a new useful method to study mineral fouling, as it gives invaluable information on the suspended particles in real time. Combining it with EIS gives complementary information on mineral accumulation on the membrane surface. Insight from this study and further use of these methods can guide the development of future strategies towards higher effluent recovery by membrane filtration.

22b - Molecular dynamics study on membrane fouling by oppositely charged proteins (V)

Jia Wei Chew (Nanyang)*

Membrane fouling continues to hamper the performance of membrane-filtration processes. A challenge with macromolecular foulants like proteins is that macroscopic characterizations, like net electrical charge, may be poorly correlated with membrane fouling. This necessitates a molecular-scale analysis of the local interactions. In this study, molecular dynamics simulations have been performed to understand the interactions between two similar-sized proteins with opposite overall charges (namely, lysozyme and α-lactalbumin) and a negative-charged membrane. Surprisingly, the protein-membrane distances and adsorption probabilities of both proteins are similar. Compared to the positive-charged lysozyme, the negative-charged α-lactalbumin exhibits (i) greater protein-membrane attractive interaction energy due to synergy among adsorption sites; (ii) lower root-mean-squared deviations (RMSD); and (iii) greater number of residues that show low root-mean-squared fluctuations (RMSF). These results indicate that local interactions are critical and thus highlight the pitfall of using the overall protein characteristics as predictors of membrane fouling.

22c - Characterization of Membrane Fouling by Zeta Potential

Vidumin Dahanayake (Anton Paar USA)* ; Stephen Hussey (Anton Paar USA); Thomas Luxbacher (Anton Paar GmbH)

Use of polymer membranes for purification and separation processes has increased steadily over the last decade. For pressure driven processes such as nanofiltration and reverse osmosis, the deposition of matter, termed fouling, presents a challenge. Traditional characterization techniques typically apply to progressive fouling but fail to analyze deposited layers. Zeta potential is an interfacial property that describes the charging behavior of surfaces. As such, the zeta potential at the interface of the membrane surface and aqueous solution explains the fouling behavior and aids in proper optimization of the surface chemistry of the polymer material.

Herein we discuss the zeta potential changes of membranes with different antifouling coatings when exposed to wastewater, humic acid, fermentation broth, and sea water. The cleaning efficiency of these membranes is also discussed as it pertains to the zeta potential.

Contact angle measurements are widely used to assess the wettability of material surfaces. Zeta potential and electro-kinetic effects are a valuable complimentary technique to contact angle for characterizing the surface functional groups that trigger wettability. At high pH, zeta potential provides information on relative hydrophobic-hydrophilic behavior emanating from solid-solute interactions.

Lastly, we will also discuss the zeta potential of particles in filtration media and how electrostatic attraction or repulsion helps to optimize membrane surfaces. Determination of membrane pore sizes using capillary flow porometry will also be described.

22d - Non-invasive and real-time monitoring of membrane fouling and cleaning

Iliane Rafaniello (SURPHASE)* ; Thomas Schäfer (POLYMAT, University of the Basque Country)

Membrane fouling can seriously affect membrane filtration processes. As fouling cannot be completely avoided, focus lies on optimizing membrane cleaning intervals in order to increase membrane lifetime while reducing process downtimes and cleaning-costs. Interviews with some of the major Spanish desalination operators revealed that until today operators lack a method sufficiently sensitive for detecting membrane fouling at an early-stage which would avoid significant drops in water flux and permit optimizing process operating conditions such as membrane cleaning protocols. Several methods have been proposed for membrane fouling detection but until today none has proven to be sensitive enough to handle membrane fouling in an anticipative way.

This work presents an approach of overcoming the limitations of conventional methods by combining two advanced surface-sensitive techniques: quartz crystal microbalance with dissipation monitoring (QCM-D) and multi-parameter surface plasmon resonance (MP-SPR). The resulting device yielded an accurate, fast, and non-invasive fouling monitoring of unprecedented sensitivity.

Results of an combined MP-SPR/QCM-D real-time monitoring device integrated into a RO system will be presented. We could successfully demonstrate that QCM-D/MP-SPR enables characterizing the build-up of even the very first fouling layers at the nanoscale and therefore possessed a sensitivity high enough such as to anticipate the need of membrane cleaning before the filtrated water flux decreases. QCM-D/MP-SPR furthermore allowed to test different membrane cleaning strategies in a time-efficient and material-extensive manner, maximizing in this way cleaning efficiency and prolonging membrane lifetime.

The implementation of the QCM-D/MP-SPR-based monitoring device on a industrial scale will be evaluated, and its potential to increase the process efficiency and mitigate membrane fouling will be critically discussed.

23a - Open-Access Database for Water Purification and Desalination Membranes

Cody Ritt (Yale University)* ; Timothee Stassin (KU Leuven); Douglas Davenport (KU Leuven); Ryan DuChanois (Yale University); Ines Nulens (KU Leuven); Zhe Yang (University of Hong Kong); Naama Segev-Mark (Technion - Israel Institute of Technology); Adi Ben

Since the advent of thin film composite polyamide membranes over 50 years ago, the permselectivity of water purification and desalination membranes has improved only marginally. The slow progression is partly attributable to limited understanding of membrane synthesis–structure–performance relationships. A centralized archive of membrane characterization data may lead to a shared understanding of features that maximize performance and unify research efforts. The Open Membrane Database, which launches online in late 2021, is a database of over 600 water purification and desalination membranes that are sourced from peer-reviewed journals, patents, and commercial product data. Here, we outline the detailed functionality of the database and the transport theory underlying the performance calculations. The open-access database may be used to benchmark novel membranes against the state of the art, conduct meta-analyses, and develop synthesis–structure–performance relationships, each of which will be critical to advancing membrane development.

23b - Scale-up of High Performance Surface Nano-Structured Reverse Osmosis TFC Membranes

Yian Chen (UCLA); Soomin Kim (UCLA); Yeunha Kim (University of California, Los Angeles); Yoram Cohen (Professor)*

Reverse osmosis (RO) membrane surface nano-structured (SNS) polyamide (PA) membranes with tethered polyacrylic acid (PAA) chains were synthesized via atmospheric pressure plasma-induced graft polymerization (APPIGP). Laboratory-scale SNS-PAA-PA RO membrane performance tuning, with respect to hydraulic permeability and salt rejection, was shown to overcome the typical perm-selectivity tradeoff. Moreover, SNS-PAA-PA membrane coupons tests demonstrated ~5-11% and ~5-7% lower flux decline, evaluated at the same initial flux, in fouling tests with aqueous BSA and sodium alginate solutions, respectively, relative to typical commercial seawater and brackish RO membranes. Moreover, excellent SNS-PAA-PA membrane cleaning, via simple DI water flush, resulted in 96 -100% permeability recovery which was greater relative to cleaning efficacy attained for tested commercial membranes. Although small laboratory scale (i.e., membrane coupons) SNS-PAA-PA membranes of high performance were synthesized via APPIGP, scalability of the process has been a challenge. Accordingly, in the current work, an initial proof-of-concept scale up was achieved using a robotic plasma surface activation system, followed by graft polymerization in a shallow reactor. SNS-PAA-PA membrane sheets, synthesized for fabrication of 2.5" x 21" length spiral-wound membrane elements, demonstrated water permeability and salt rejection ranges comparable to the Dow SW30 membrane used for a baseline performance comparison. Fouling tests with BSA and sodium alginate solutions demonstrated 8% and 12% lower flux decline, respectively, relative to the Dow SW30 element. Simple DI water cleaning of the fouled SNS-PAA-PA membrane elements provided 100% permeability recovery relative to 92% for the Dow SW30 element. The present study suggests that the APPIGP process can be scaled up, e.g., via continuous surface activation, for fabrication of large commercial-size high performance spiral-wound SNS-PAA-PA RO elements.

23c - Recovering end-of-life reverse osmosis membrane productivity using chlorination

Bianca M. Souza Chaves (The University of Arizona)* ; Mohammed Alhussaini (The university of Arizona); Walter Betancourt (The University of Arizona); Andrea Achilli (The University of Arizona)

The expected life of reverse osmosis (RO) membranes is still a critical challenge due to irreversible fouling that reduces their productivity. End-of-life RO membranes are periodically replaced and disposed of in landfill. Extending the usable life of RO membranes with chlorination can increase equitable access to reliable water supply. The goal is to restore water permeability of fouled membranes from an engineering-scale RO system treating reclaimed water, and gain a comprehensive understanding on the changes in solute permeation due to chlorination. The membrane was oxidized by progressively exposing the elements to 2,000 ppm-h of chlorine for 5 times after chemical cleaning. Water/conductivity/ion/organic permeability coefficients and rejection were calculated and related to the chlorine dose. Six natural occurring viruses differing in size were also monitored. After 5 chlorine doses, the water permeability was recovered to the typical range of the system (~1.3 L m-2 h-1 bar-1), a 3-fold increase compared to the end-of-life membranes. Interestingly, conductivity and ion permeability slightly decreased after the first and second dose, likely because the chlorine removed part of the irreversible fouling and thus reduced concentration polarization. After 3 doses, as the membrane surface oxidized, a general increase of ion permeability was noted, where more monovalent ions permeated through the membrane, while divalent ions rejection remained high and constant. Similarly, the fluorescence intensity decreased before 2 doses in RO permeate, followed by an increase after 3 doses, and only low molecular weight substances (< 1 kDa) were detected. All viruses were below or near the limit of detection, indicating that the oxidation has not compromised the integrity of the membrane or its ability to remove virus-sized particles. Results suggest that targeted chlorination may extend RO membranes lifespan by increasing water permeability while maintaining RO-like selectivity.

23d - Artificial Water Channels- toward Biomimetic Membranes for Desalination

Mihail BARBOIU (Institut Europeen des Membranes)*

This lecture discusses the incipient development of the first artificial water channels systems. We include only systems that integrate synthetic elements in their water selective translocation unit. Therefore, we exclude peptide channels because their sequences derive from the proteins in natural channels. We review many of the natural systems involved in water and related proton transport processes. We describe how these systems can fit within our primary goal of maintaining natural function within bio-assisted artificial systems. In the last part, we present several inspiring breakthroughs from the last decade in the field of biomimetic artificial water channels. All these examples demonstrate how the novel interactive water-channels can parallel biomolecular systems. At the same time these simpler artificial water channels offer a means of understanding water structures useful to understanding many biological scenarios. Moreover they can be used for the preparation of highly selective membranes for desalination.

1] M. Barboiu and A. Gilles, Acc. Chem. Res. 2013, 46, 2814–2823.

[2] M. Barboiu, Angew. Chem. Int. Ed. 2012, 51, 11674-11676.

[3] Y. Le Duc, et al., Angew. Chem. Int. Ed. 2011, 50(48), 11366-11372.

[4] E. Licsandru, et al., J. Am. Chem. Soc., 2016, 138, 5403-5409.

[5] M. Barboiu, Chem. Commun., 2016, 52, 5657- 5665.

[6] I. Kocsis, et al. Science Adv. 2018, 4, eaao5603.

[7] M. Di Vincenzo, A. Tiraferri, V.-E. Musteata, S.Chisca, R.Sougrat, L.-B. Huang, S. P. Nunes, M. Barboiu, Nat. Nanotechnol. 2021, 16, 190-196.

[8] M. Di Vincenzo, et al. Proc. Natl. Acad. Sci. USA, 2021, DOI : 10.1073/pnas.2022200118

24a - Water and organic co-transport in carbon molecular sieve membranes

Young Hee Yoon (Georgia Institute of Technology)* ; Ryan Lively (Georgia Tech)

Understanding molecular transport of mixtures in confined microporous spaces is an important research area for a variety of membrane separation processes. One important example is the transport of water and organic mixtures, especially in conditions of high guest loadings. Separation of water and small-molecule organic solvents is a challenge in industrial applications such as purification of produced water, solvent recovery from a water-organic mixture, and water removal from black liquor. Carbon molecular sieve (CMS) materials are a potential candidate for scalable and high-performance reverse osmosis membranes due to impressive chemical and thermal stabilities. The amorphous nature of the CMS microstructure results in significant difficulties in the use of computational modeling of molecular transport processes. Therefore, an experimental and mechanistic study of water and organic solvent transport in CMS membranes is pursued here. In this work, we investigated the sorption, diffusion, and permeation behavior of organic compounds and water in polyvinylidene fluoride (PVDF) derived CMS (PVDF-CMS). The comparison shows how the properties of penetrants such as hydrophilicity and molecular size of permeate affect the transport behavior in CMS. Basic transport and sorption parameters are utilized to estimate how water-organic mixtures will transport in a CMS membrane. Using the generalized Maxwell-Stefan equation and ideal adsorbed solution theory (IAST) based on pure component isotherms, flux equations of water and p-xylene mixture in PVDF-CMS membranes are developed and compared with the experimental data.

24b - Designing Few-nanometer organosilica membranes with hydrothermal stability for selective hydrogen separation

Thien N Tran (University at Buffalo, The State University of New York)* ; Liang Huang (The State University of New York at Buffalo); Vinh Bui (University at Buffalo); Lingxiang Zhu (U.S. Department of Energy National Energy and Technology Laboratory); Haiq

Nanoporous silica membranes exhibit excellent H_2/CO_2 separation properties for sustainable H_2 production and CO_2 capture but are conventionally prepared via complicated thermal processes above 400 oC, inhibiting their scalable production at low cost. Here, we demonstrate the rapid fabrication (within 2 minutes) of ultrathin silica membranes (≈6 nm) via oxygen plasma treatment of polydimethylsiloxane based thin-film composite membranes at 20 oC. The effect of plasma treatment conditions such as exposure time, oxygen concentration on the structure-properties relationship of the organosilica layer was systematically investigated. The resulting organosilica membranes exhibit H_2 permeance of 280 - 930 GPU (1 GPU = 3.347×10-10 mol m-2 s-1 Pa-1) and H_2/CO_2 selectivity of 93 - 32 at 200 °C, far surpassing state-of-the-art membranes and Robeson's upper bound for H_2/CO_2 separation. When challenged with a 3-day simulated syngas test containing water vapor at 200 °C and a 340-day stability test, the membrane shows durable separation performance and excellent hydrothermal stability. The robust H_2/CO_2 separation properties coupled with excellent scalability demonstrate the great potential of these organosilica membranes for economic H_2 production with minimal carbon emissions.

24c - Development and fabrication of high-performance Pd-based CMRs for ammonia decomposition

Rok Sitar (Colorado School of Mines)* ; Javishk Shah (CSM); J. Douglas Way (CSM); Colin Wolden (Colorado School of Mines)

Catalytic membrane reactors (CMRs) can be used to simultaneously decompose ammonia and purify the resulting hydrogen, making this technology attractive for supplying hydrogen-fueling stations. Reproducibility of CMR fabrication and the durability of the CMRs is essential for the wide-scale deployment of such technology. We have developed high-performance CMRs based on asymmetric tubular YSZ support where the Ru catalyst was impregnated in the pores of the support and serves as active sites for the Pd nucleation. A critical step in the CMR fabrication is the deposition of the Pd membrane, and in this talk, we will describe an improved electroless plating setup that includes substrate rotation, ultrasonic agitation, and precise temperature control equipment. Through appropriate control of these parameters, the membrane thickness required for CMR applications was reduced from 6.4 to less than 4 microns while maintaining extremely low N2 leak rates of <4x10-4 mol/(m^2 s) at ?P = 75 psi. This is attributed to the displacement of the gas bubbles formed during the reduction and Pd plating steps, which mitigates pinhole formation. Reproducibility was also significantly improved, with 10 sequentially fabricated membranes displaying nominally equivalent performance. The durability of the improved Pd-based CMR was evaluated under ammonia flow at operating temperatures of 425 – 450°C for over 500 hrs with a single complete system shutdown. All aspects of CMR performance, including ammonia conversion and hydrogen recovery, significantly improved while operating in this reducing environment. This was attributed to improved H2 permeance which increased by 28% over the course of testing, and equally important, with no loss of selectivity. We expect that the improved electroless plating techniques described herein will be equally important for the synthesis of Pd alloy membranes, particularly where composition control is important.

24d - Designing graphene oxide membranes by etching and polydopamine intercalation for high-efficiency dye removal

Xiaoyi Chen (State University of New York at Buffalo); Erda Deng (State University of New York at Buffalo); Xiaoci Lin (State University of New York at Buffalo); Lingxiang Zhu (U.S. Department of Energy National Energy and Technology Laboratory); Haiqing

Graphene oxide (GO) has been extensively investigated for water purification owing to adjustable sub-nm channel sizes with molecular-sieving ability. However, its good dispersibility in water leads to poor durability, while conventional approaches of reduction and cross-linking improve the durability but decrease water permeance. Herein, we demonstrate an effective approach to enhance water permeance while maintaining durability by chemical etching to create holey GO (HGO) nanosheets and intercalation with polydopamine (PDA) nanoparticles. Specifically, GO nanosheets were chemically etched for 2 or 4 h using H_2O_2 to obtain HGO2h or HGO4h, which were then co-deposited with dopamine on porous supports. The prepared membranes were thoroughly characterized using SEM, AFM, and XPS. The nanopores created on the HGO shorten the diffusion path, increasing the water permeance, while the PDA intercalation changes the channel sizes and increases durability. For example, PDA/GO membrane with a thickness of 20 nm shows water permeance of 46 LMH/bar, while the PDA/HGO4h with a thickness of 37 nm shows a water permeance of 274 LMH/bar. Additionally, centrifugation treatment of HGO nanosheets increases the dye rejection of the PDA/HGO membranes, presumably because the removal of small pieces of HGO nanosheets contributes to a more regular stacking of HGO. The optimized membrane with 18 nm PDA/HGO2h exhibits water permeance of 119 LMH/bar and a Congo Red rejection of 98.5%, better than many GO membranes reported in the literature. Moreover, the membrane exhibits stable performance in a 3-day continuous cross-flow filtration test, indicating its durability and potential for practical applications.

25a - High-capacity adsorbents with hierarchical structures printed from polymer composites

Jialing Xu (University of Notre Dame); Thomas Kasl (University of Notre Dame); Kevin Gabriel Alvarez (University of Notre Dame); Bill Phillip (Notre Dame)*

The increasing global population, improved standards of living, and the expansion of irrigated agriculture will make meeting the demand for fresh water a challenge. Selective adsorbents that target the capture of contaminants are one technology for helping to ensure a supply of water that is fit for its intended use. Adsorbent design has focused largely on the development of novel chemistries. However, while there is scope for improvements in materials properties, process analysis suggests that corresponding efforts need to be dedicated to addressing limitations at the device-scale for the role of adsorbents to be expanded. In this talk, we will discuss how additive manufacturing and membrane fabrication methods (e.g., vapor induced phase separation) can be combined to fabricate sorbents with hierarchical structures. Solutions containing polysulfone and a polystyrene-block-poly(acrylic acid) (PS-PAA) block polymer were deposited as the filaments of structured sorbents utilizing solvent-assisted three-dimensional printing. Through the appropriate selection and control of the printing environment, the filaments were manufactured to possesses a bi-continuous network of pores ~500 nm in diameter. The PAA moieties of the block polymer, which segregate to the nanopore walls during fabrication, allowed the chemistry to be tailored for heavy metal removal via solid state coupling reactions. Here, terpyridine groups were incorporated along the pore wall to develop a sorbent capable of efficiently removing metal ions under conditions where the contaminant is present at trace concentrations (e.g., < 10 ppm). This molecular design enables the highly efficient purification of simulated ground water or seawater solutions by capturing 99+% of the metal ions dissolved in them. At the device-scale, the hierarchical structures of the sorbent reduce mass transfer limitations that hinder the performance of traditional resins.

25b - The Future Use of Atomic Layer Processing in Membrane Production

David S Bergsman (University of Washington)*

Recent years have seen a surge of interest into improving the scalability of atomic layer processing (ALP) tools, such as work in spatial atomic layer deposition (ALD), molecular layer deposition (MLD), and vapor-phase infiltration (VPI). This improved scalability, along with a continued decrease in the cost of these processes, has allowed these tools to be considered outside the semiconductor industry, such as in membrane synthesis. Because membranes rely upon precise materials properties and structure to operate effectively (e.g. narrow pore size distributions, non-fouling surface chemistries, and robust rheological properties), they are an excellent candidate for the use of atomic layer processing tools in their synthesis.

This presentation will highlight the potential use of atomic layer processing tools for membrane separations, reviewing recent advancements in this area. Generally, these tools can either be used to enhance an already existing material, such as by coating a porous structure or infiltrating a bulk material, or they can be used to create the selective layer itself, through the deposition of a thin selective layer. First, the use of ALD, MLD, and VPI to enhance membrane performance will be highlighted, including work to decorate membrane surfaces, control pore sizes, and infiltrate polymers to impart conductivity or chemical resilience. Then, cases where ALD and MLD have been used to create the active separating layer will be discussed, such as in ALD-based inorganic ion conductors, MLD-based desalination membranes, and hybrid ALD/MLD-based porous layers. Finally, promising future directions and the improvements that will ultimately be necessary for these tools to be used in commercial membrane synthesis will be explored.

25c - Customized thin film composite membranes using additive manufacturing

Xin Qian (University of Connecticut)* ; Tulasi Ravindran (University of Connecticut); Maqsud Chowdhury (University of Connecticut); Samuel J Lounder (Tufts); Ayse Asatekin (Tufts); Rhea Verbeke (KU Leuven); Ivo Vankelecom (KU Leuven); Jeffrey R McCutcheon

We proposed a novel additive manufacturing method to make thickness-controllable thin film composite membranes (TFCs) from a variety of polymers (even those not traditionally considered for TFCs). Conventional membrane fabrication techniques, such as phase inversion and interfacial polymerization, lack resolution for thickness and roughness control and may limit which materials can be used for membranes when a sub-100 nm film is desired. Our novel additive manufacturing technique employs an electrospray 3D printer that precisely deposits materials on a membrane substrate to achieve sub-5 nm thickness resolution and produces membranes with exceptional permeance and selectivity. In our work, we have demonstrated that electrospray additive manufacturing achieves the highest resolution (for thickness control) among all other 3D printing techniques considered for membrane production. Due to its advantageous printing resolution and wide material compatibility, our approach has enabled us to print multiple membrane materials for a variety of applications, including polyamide reverse osmosis (RO) membranes, zwitterionic copolymer nanofiltration (NF) membranes with exceptional permeance, and polyepoxyether NF membrane with robust chlorine tolerance.

25d - Antimicrobial Facemask Membrane Filters Cast on a 3D Printed Support

Ebuka Ogbuoji (University of Kentucky)* ; Isabel Escobar (University of Kentucky)

The Covid 19 pandemic has led to growing demands for personal protective equipment (PPE) to effectively control the spread of the virus. Facemasks are an effective defense against aerosols containing pathogenic bacteria and viruses such as Sars-Cov-2. Membrane filters have been used extensively in face masks to remove these microbes from the air. These filters are usually designed for single-use due to inadequate and laborious cleaning/decontamination techniques. This work attempts to make a breathable antiviral face mask by immobilizing silver nanoparticles (AgNPs), which could suppress bacterial and viral activity on a cellulose acetate (CA) membrane filter required for mask production. AgNp was chemically immobilized by attaching a polymerized epoxy, glycidyl methacrylate (GMA) to CA, allowing for more functionalization of the CA/GMA copolymer. Cysteamine was then combined with the CA/GMA complex, providing thiol groups that immobilized AgNP's on the membrane surface. FTIR analysis confirmed the successful polymerization of the monoGMA, while electron microscopy and X-ray energy dispersive spectroscopy was used to verify the presence of silver on the CA membrane surface. The resulting membrane filters are quite thin and require support for use in mask production. We have used a 3D printed support to ensure strong membrane filters for mask production in this work. An airflow test was carried out on the unmodified CA membrane on a 3D support to ensure breathability. A high airflow resistance was observed through the membrane at pressures up to 10psi. This was hypothesized to be due to small pore sizes inherent in biobased membranes. Therefore, polyethylene glycol (PEG), an organic chemical known to form pores in membranes, was introduced in the dope solutions, and subsequent increases in pore sizes and air permeability were observed.

26a - Chemically Resilient Hollow Fiber Nanofiltration Membranes Fabricated from Copolymers for Organic Solvent Nanofiltration

Michael Dugas (University of Notre Dame)* ; Shukun Zhong (University of Notre Dame); Bill Phillip (Notre Dame)

The need for membrane platforms to accommodate the highly varied demands for organic solvent (OS) separations has emerged to meet the demand for energy-efficient separations. Nanofiltration membranes based on copolymer materials are a promising platform because they can be engineered at the molecular scale to address an array of process needs. In addition, these copolymer materials can be reacted with functionalities that can crosslink and create an electrostatic charge, thereby creating a durable, more effective membrane. Here, a resilient membrane is developed through the molecular design of a poly(trifluoroethyl methacrylate-co-oligo(ethylene glycol) methyl ether methacrylate-co-glycidyl methacrylate) copolymer that can be dip-coated onto hollow fiber supports. The epoxide rings on the copolymer were functionalized post-casting with diamines of various chains lengths to incorporate positively-charged moieties along the pore walls that crosslink the overall polymeric structure. By varying the chain length, XPS analysis has shown that the extent of crosslinking is affected. This extent of crosslinking in turn affects the pore diameter of the membrane in various OS environments, including THF, DMF, and EtOH without damaging the structure of the membrane itself. Preliminary results also demonstrate that the membranes can separate based not only on size-exclusion principles, but also on the charge of the solute, as seen by an over 90% rejection of Martius Yellow in EtOH. Finally, the copolymer crosslinked with the diamines resulted in membranes that could operate continuously in EtOH as well as cycled between EtOH and water over a prolonged period with little to no degradation of the membrane. These results demonstrate that by combining the versatility of molecularly designed copolymers with functionalities that can be utilized to increase the integrity of the membrane, copolymeric membranes can further the use of membrane technologies into emerging OS applications.

26b - The role of skin layer pores in organic solvent reverse osmosis permeation through polymer membranes

Hye Youn Y Jang (Georgia Institute of technology)* ; Ryan Lively (Georgia Tech)

The fractionation of complex liquid hydrocarbon mixtures is an important emerging area of membrane science. Polymeric asymmetric hollow fiber membranes can enable this concept, especially if the size and number of defects in the membrane skin layer can be precisely engineered. Here, we fabricated various "defect-engineered" hollow fiber membranes by modifying the composition of the volatile solvents of the polymeric solution as well as spinning parameters to improve the organic solvent reverse osmosis (OSRO) separation performance of complex aromatic hydrocarbon mixtures. The quality of the membranes was probed using both gas and liquid permeation tests, which provided the data for a useful fundamental analysis that estimates the pore size and surface porosity of the asymmetric hollow fiber membranes. That information, combined with organic solvent permeation data, provides a workable platform to estimate the relative permeation rates of organic molecules within a complex mixture. For instance, we will assess the ability of combined transport modality permeation models (i.e., sorption-diffusion + hydraulic flow) to predict the membrane's ability to fractionate complex mixtures of aromatic hydrocarbons. Useful approximations and apparent deficiencies in this transport "framework" will be discussed. Although we have employed hydrocarbon mixture fractionation to illustrate our approach, we believe the membrane-based fractionation concept to be highly relevant across a range of existing and emerging separation challenges, including biorefinery separations and downstream separations in futuristic "e-refineries"

26c - All Organic Thin-Film Nanocomposite Membranes for Organic Solvent Filtration

Syed Ibrahim Gnani Peer Mohamed (University of Nebraska-Lincoln); Siamak Nejati (University of Nebraska-Lincoln)* ; Mona Bavarian (University of Nebraska-Lincoln)

The incorporation of nanomaterials in thin-film composite (TFC) membrane for organic solvent nanofiltration (OSN) has proved as a practical approach to fabricate high-performance membranes. Herein, we report on the synthesis of poly(divinylbenzene-co-perfluorodecyl acrylate) (poly(DVB-co-PFDA)) nanoparticles (NPs), via free radical precipitation polymerization, that were further integrated into the architecture of the TFC membrane to construct thin-film nanocomposite (TFN) membranes. The TFN membrane prepared with poly(DVB-co-PFDA) NPs in the PA layer exhibited toluene permeance of 0.6 L/m2 h bar, which is twice more than that of the control TFC membrane (0.3 L/m2 h bar)–without compromising the solute rejection. Furthermore, the fabricated membrane exhibited higher methanol permeance of 5.4 L/m2 h bar with a molecular weight cut-off (MWCO) of ~300 Da. The integration of hydrophobic fluoro-containing polymeric NPs in the fabrication of TFN OSN membrane provides a promising path to more robust TFN membranes with improved performance.

26d - Separation of organic solvents using Dual-Layer Hollow Fiber Mixed Matrix Membranes

Conrad J Roos (Georgia Institute of Technology)* ; Hye Youn Y Jang (Georgia Institute of technology); Dylan Weber (Georgia Institute of Technology); Ryan Lively (Georgia Tech)

Membrane-based organic solvent separations hold the potential to reduce the energy and carbon footprint of existing and emerging processing industries. High-performance membranes will accelerate this transition, especially if the membrane fabrication can be done in a scalable manner to match the processing requirements across the full range of industries. This work includes a techno-economic analysis of xylene selective carbon molecular sieve membranes fabricated previously in our group. This analysis demonstrates that these fabricated membranes are competitive with commercial adsorption processes for the purification of xylene isomers on a cost basis. This work expands beyond the techno-economic analysis's focus on xylene isomers with the successful fabrication of mixed matrix hollow fiber membranes for organic solvent separations. The benefits of mixed matrix membranes have been well documented, and the hollow fiber morphology of these membranes presents a scalable and commercially attractive membrane morphology. The polyimide Matrimid®and the metal organic framework UiO-66 were paired together for this membrane. The final mixed matrix hollow fiber membranes were tested using an organic solvent mixture of toluene and mesitylene, a boost in permeance due to UiO-66 inclusion without a drop-off in rejection. This work demonstrates the successful dual-layer fabrication and post-treatment of mixed matrix hollow fiber membranes for organic solvent reverse osmosis applications.

27a - Conversion of Food Waste to Levulinic Acid Using A Catalytic Membrane Reactor

Zhexi Zhu (University of Arkansas)* ; Davar Sasongko (SIEV Technologies); Ranil Wickramasinghe (University of Arkansas); Xianghong Qian (University of Arkansas)

One-pot conversion of biomass to levulinic acid (LA), a promising platform chemical of bio-refinery, is an important technology in the biomass conversion process. The production yield is critical to the competitiveness of biomass conversion, meanwhile mild reaction condition is important to industrial operation. Here, cellulose as well as food wastes from vegetables and other starch-based food materials have been investigated for their conversion to levulinic acid using a unique solid acid catalyst immobilized on a membrane substrate. Our reusable membrane catalyst is superior to the corrosive homogeneous acid or toxic metal-based catalyst. In addition, the membrane substrate enables the immediate separation of the levulinic acid from the rest of the feed stream driving the reaction to completion and improving the yield. A levulinic acid yield of over 80% has been achieved for the model cellulose compound. For starch-based food waste such as rice and noodles, over 90% of levulinic acid yield was obtained. For the more recalcitrant lignocellulosic vegetables, a 50% levulinic acid yield has also been achieved indicating the promising potential of our technology for food waste utilization.

27b - Membrane Applications in Dairy Industry – Status, Challenges and Opportunities for Innovation

Bing Liu (Leprino Foods)*

Membrane technology has been widely used in the dairy industry and become an essential technology for producing a variety of dairy products, such as whey protein concentrate (WPC), whey protein isolate (WPI), milk protein concentrate (MPC), micellar casein concentrate (MCC) and native whey protein concentrate (NWPC) etc. Compared to conventional technologies, the membrane technology is more advantageous. There still be many potential opportunities for new membrane applications supporting the development of new dairy products. Among the membrane technologies, the RO is used for de-water and water reuse. The NF is used for further concentration of whey protein and demineralization. The UF is used for protein concentration and the MF is used for defatting and casein separation. Due to the nature of dairy feed streams the applications of membrane technology to the dairy industry are still facing many challenges, such as flux decline by concentration polarization or membrane fouling by fat/protein which results in using more membrane surface area and high capital investment; blistering appeared on glue seam which results in sanitary issue and chorine attack on membranes (from daily CIP) which results in short membrane life etc. Although lot of progresses made to improve membrane performance (such as better protein capture rate, more surface area, better blistering resistance etc.), no breakthroughs have been made in the recent years. With the rapid evolution of new membrane technologies, it provides great opportunities for potential innovations for dairy membrane applications. Innovations on any of above challenges could have significant impacts either financially or on energy savings for the dairy industry. This presentation will provide an introduction of membrane applications in the dairy industry as well as the status of current dairy membrane processes. It will also discuss the challenges in dairy membrane applications as well as opportunities for innovation.

27c - Micellar casein from microfiltration of skim milk and its beverage application (V)

Ni Cheng (Milk Specialties Global)* ; MaryAnne Drake (North Carolina State University); Dave Barbano (Cornell University)

Evolving with the membrane technology and the consumer desire for clean label, nutritional and functional protein ingredients for dairy beverages, fractionation of milk protein by microfiltration opens a new avenue for dairy protein ingredient innovations. Microfiltration of milk, pasteurized skim milk or milk protein concentrate after ultrafiltration, can produce various purity of micellar casein and milk driven whey protein. Micellar casein is a good carrier for protein and calcium fortification for beverages and heat stable comparing with whey protein. Understanding the inclusion amount of micellar casein concentration or casein to true protein ratio and its impacts on sensory and physical properties of dairy protein beverages is needed. A 95% milk driven whey protein removed micellar casein concentrate was produced from pasteurized skim milk by using a 3x, 3 stages ceramic microfiltration process at 50°C while milk driven whey protein isolate was further processed through 2 stages ultrafiltration from the microfiltration permeate. 60 milk beverages were formulated with fat level from skim, 1% to 2%, casein to true protein ratio from 5%, 25%, 50%, 75% to 80% and 4 true protein levels from 3.00%, 3.67%, 4.34% to 5.00%, and a constant lactose of 4.65%. All formulations were homogenized and pasteurized, and then profiled trained sensory descriptive panel (appearance, flavor and texture) and evaluated for physical properties (such as, color and viscosity). Casein to true protein ratio had a large impact on beverages whiteness and yellowness, with higher the ratio, the more white and less yellow. Higher casein concentration increased instrumental viscosity and decreased flavor intensities while less casein concentration increased aroma intensity, sweet aromatic, cooked/sulfur and cardboard/doughy flavors, and sensory yellowness within each fat level and across fat levels.

27d - Membranes for the removal of ethylene from produce ripening environments

Hannah Murnen (Compact Membrane Systems); Christine Parrish (Compact Membrane Systems)* ; Sudip Majumdar (Compact Membrane Systems); Stuart Nemser (Compact Membrane System); Gary Ward (StePac); Ohad Nerya (StePac); Deirdre Holcroft (Holcroft Postharvest Co

Post-harvest technologies for fruits and vegetables, including refrigeration, modified atmospheric packaging (MAP), controlled atmospheres and ethylene management, have led to significant increases in shelf life and improvements in quality. Ethylene, a hormone generated by post-harvest produce, leads to premature ripening and rapid decay. Techniques to date that are used to manage/destroy ethylene include sorption, chemicals (1-methylcyclopropene), and reactions to modify ethylene.

This paper will discuss development of ethylene-selective membranes for removing ethylene from produce environments, and in the process, increasing post-harvest life. Compact Membrane Systems (CMS) has developed membranes with very high ethylene permeance (e.g., 500 GPU) and very low oxygen and nitrogen permeance (1 GPU). These membranes work well at low ethylene concentrations (e.g., <1.0 ppm) typically measured in produce packaging and controlled atmosphere environments.

Membrane performance related to ethylene removal has been evaluated in multiple modes of post-harvest produce operations. This includes modified atmosphere packaging (MAP) where gases are stagnant in a 5-10 pound bag of fruits and vegetables, and also controlled atmospheres where large volumes of refrigerated gases are circulated in the containers. Data taken by our partner StePac, a packaging company located in Israel, demonstrates the effectiveness of CMS membranes in extending the shelf-life of bananas and kiwifruits.

Potential operational designs for both packaging and controlled atmosphere will be discussed. Representative systems will be presented to demonstrate the tradeoff of membrane area requirements versus projected ethylene levels (e.g., 0.1, 0.05 ppm).

28a - Micro- and nano-patterned polymer electrolyte membranes for electrochemical energy conversion

Christopher G Arges (Louisiana State University)* ; Subarna Kole (LSU Chemical Engineering Department)

High-surface area interfaces in polymer electrolyte membranes can enhance charge-transfer kinetics in membrane electrode assemblies for fuel cells and promote water-dissociation kinetics in bipolar membranes. In the context of bipolar membranes, we have systematically controlled the interfacial area in the junction region via soft-lithography. Increasing the interfacial area in the junction region by 2.2x reduces the onset potential for water-splitting by over 250 mV and enhances the current density for water-splitting. The lateral feature sizes generated from conventional soft-lithography using PDMS molds is limited to about 1 μm. Hence, generating greater interfacial area values on polymer electrolyte membranes via smaller periodic feature sizes over a large area without direct writing (i.e., ebeam lithography) necessitates an alternative, low-cost and scalable nanopatterning platform. In this talk, we'll show block copolymer lithography as a scalable, nanopatterning scheme to fabricate nanopatterned polymer electrolyte membrane surfaces with lateral, periodic feature sizes below 100 nm.

28b - Controlling the Structure and Rotation Direction of Electroconvection by Membrane Surface Modification

Felix Stockmeier (DWI - Leibniz Institute for Interactive Materials)* ; Daniel Felder (DWI - Leibniz Institute for Interactive Materials); Michael Schatz (DWI - Leibniz Institute for Interactive Materials); Malte Habermann (DWI - Leibniz Institute for Inte

A decade ago, microscopic two-dimensional flow visualization proved the theoretically predicted existence of electroconvection (EC) as well as its decisive role in the destabilization of the concentration polarization layer at ion-selective interfaces, like ion exchange membranes or electrodes. EC induces ion transfer as it convectively transports ions into the ion-depleted diffusion layer next to said interfaces. Thereby, the increased ion concentration at the interface allows for an overlimiting current density. Several studies focus on understanding and controlling the hydrodynamics of the EC vortex field to extend the operational range of electrically driven membrane processes towards overlimiting current densities.

We proposed a membrane surface modification method that has shown promising mass transfer improvements. The modification introduces a surface charge inhomogeneity by precise surface patterning with microgel suspensions. The modified membranes result in an early transition to overlimiting currents with reduced resistance. However, the influence of the pattern on the system's hydrodynamics is still unexplored. We investigate the impact of such surface modifications on the EC vortex field's build-up using 3D particle tracking velocimetry. With this technique, we determine the vortex field structure, its rotation direction, and the structural stability in steady-state for modified membranes compared to a homogeneous untreated membrane surface.

We show that the microgel patterns structure the vortex field and dictate a rotation direction. Although the structure is not maintained in the steady-state, the vortex field still differs from the case of an untreated membrane. The results for a complex pattern structure confirm that this modification offers full control of the EC vortex field during its build-up. This emphasizes the potential of tailored membrane surfaces to overcome today's limitation of electrically-driven membrane processes.

28c - Influence of Electrolyte on the Concentration-induced Conductivity-permselectivity Tradeoff of Ion-exchange Membranes

Yuxuan Huang (Columbia University)* ; Hanqing Fan (Columbia University); Ngai Yin Yip (Columbia University)

Ion-exchange membranes (IEMs), which are highly charged polymeric films, are widely used in water and energy applications. Concentration-induced tradeoff relationships between conductivity and permselectivity constrain the performance of IEM-based processes. Though classic Donnan-equilibrium-based theory offers a good qualitative or semi-quantitative explanation on the origin of the phenomena, the tradeoff behaviors of IEMs in different electrolyte solutions and the influence of electrolyte on the tradeoff relationships have not been fully investigated. The knowledge gap hinders the development of IEM separation to treat mixed electrolyte solutions. In this study, we investigated the impacts of electrolytes on the concentration-induced conductivity-permselectivity tradeoff of commercial cation and anion exchange membranes. The effective ionic conductivity and apparent permselectivity of IEMs were characterized in different electrolyte solutions and concentration levels. The increasing external concentration enhanced the effective conductivity, which was mainly determined by the counterion identity and was insensitive to the co-ions. The apparent permselectivity declined with external concentration, with higher valency counterions and lower valency co-ions yielding lower permselectivity. By combining the conductivity and permselectivity data, different tradeoff behaviors caused by external concentration were observed in different electrolyte solutions. Lastly, conductivity and permselectivity models were developed to facilitate the understanding of the primary factors that affect the tradeoff relationships. Qualitative consistency between experimental and models was observed, indicating that the theoretical framework underpinning the model can explain the transport phenomena. The study advances the understanding of IEM conductivity-permselectivity relationship in different electrolytes and provides guidance for improving IEM performance in complex water chemistries.

28d - Reverse Osmosis vs. Electrodialysis: Identifying the Most Energy Efficient Technology for Brackish Water Desalination Applications

Sohum Patel (Yale University)* ; Maarten Biesheuvel (Wetsus); Menachem Elimelech (Yale University)

Global water scarcity persists as one of the greatest challenges of our time, affecting two-thirds of the world population and expanding at an alarming rate. Effectively combatting water scarcity requires the harnessing of unconventional, typically saline, water sources through the use of desalination processes. Hence, as our reliance on desalinated water grows, the identification and implementation of the most energy efficient technology is critical towards ensuring sustainable water production. In a recent study, we focused on determining the energy consumption of desalinating brackish waters (salinity range of ~1 g L-1 to 10 g L-1) with the two most prevalent technologies — reverse osmosis (RO) and electrodialysis (ED). In our analysis, we use rigorous process scale models to perform a systematic and direct comparison of each technology across a wide range of desalination conditions. Notably, by maintaining the same separation parameters (i.e., feed salinity, salt removal, water recovery) and productivity among each technology, we ensure that our comparison is valid and fair. Our results indicate that both ED and RO are capable of operating with high energy efficiency (>30%), though for starkly different conditions. Overall, our analysis clearly indicates that ED shows optimal performance for low feed salinities and relatively small extents of salt removal, whereas RO excels for the treatment of higher feed salinities (> 5 g L-1) which requires extensive degrees of water-salt separation. Specifically, we determine that ED is generally more energy efficient than RO for feed salinities < 3 g L-1, though its widespread implementation remains hindered by the high cost of ion-exchange membranes. We therefore suggest that future studies further extend on the performed energy consumption comparison with detailed technoeconomic analyses to provide a more holistic assessment of each technology.

29a - Elucidating the structure-performance relationship in commercially relevant virus filters using 3D electron tomography

Kaitlyn P Brickey (Pennsylvania State University)* ; AAA ZZZ (Pennsylvania State University); Enrique Gomez (Penn State)

Virus filtration membranes are instrumental in ensuring the safety of high value biopharmaceuticals like monoclonal antibodies but present unique challenges. The desired recovery of product (10-12 nm in size for monoclonal antibodies) must be high while providing >99.9% removal of small parvoviruses (~20 nm in size). Additionally, many studies have noted reduced performance for some virus filters under certain process conditions, highlighting the need to understand how morphology governs virus filter performance. In this study, we used FIB-SEM tomography to evaluate the 3D pore structure of 3 commercial virus filters: the asymmetric Viresolve®Pro and NFP filters and the relatively symmetric Pall Pegasus™SV4. Pore-network models were developed from the 3D structures to describe the distribution of body/throat sizes throughout virus-retentive regions. The asymmetric filters have a clear difference in body/throat size, with the throat size at the filter exit consistent with the virus removal capability of the Viresolve®Pro filter (21 nm throat size) compared to the Viresolve® NFP filter (29 nm throat size). The symmetric SV4 shows little differences in body/throat size—the throat size is larger than 20 nm, while the number of virus-retentive throats remains nearly constant through the filter. The 3D structures were also used to evaluate the porosity, tortuosity, and connectivity through the filter. Connectivity is highest for the SV4 filter and lowest for the Viresolve®NFP filter, consistent with independent experimental measurements. Both asymmetric filters decrease in connectivity near the filter exit. Finally, reconstructions were tested using simulation packages to confirm virus retention/flux and study morphology in virus capture regions. These results provide the first extensive 3D characterization of a variety of virus filters and provide a new way to quantify the complex relationship between morphology and performance in virus filtration membranes.

29b - Interaction-Based Ion Selectivity Exhibited by Self-Assembled, Cross-Linked Zwitterionic Copolymer Membranes

Samuel J Lounder (Tufts)* ; Ayse Asatekin (Tufts)

Water filtration membranes that achieve advanced ion selectivity are urgently needed for applications such as chloride/ fluoride separation, precious metal recovery, and boron (borate) removal. Current synthetic membranes, which separate solutes primarily by size and charge differences, are limited in their use for discriminating between ions of equal charge. Biological nanopores, in contrast, achieve exquisite ion selectivity due to interactive functional groups that line the highly structured pore wall. The favorability and strength of these interactions control ion partitioning and diffusion rate, providing a powerful mechanism for separating similarly sized ions. This inspires the design of synthetic membrane filters with analogous structure to achieve next-generation ion selectivity.

In this work, we combine polymer self-assembly and cross-linking to develop ion-selective membrane filters. We investigate the separation capabilities of cross-linked zwitterionic copolymer membranes, a novel self-assembled membrane system featuring sub-nanometer zwitterionic nanochannels. We demonstrate that selective zwitterion-anion interactions simultaneously control salt partitioning and diffusivity, with the permeabilities of $NaClO_4$, NaI, NaBr, NaCl, NaF, and Na_2SO_4 spanning roughly 3 orders of magnitude over a wide range of feed concentrations. We model salt flux using a one-dimensional transport model based on the Maxwell-Stefan equations, and show that diffusion is the dominant mode of transport for 1:1 sodium salts. Differences in zwitterion-Cl^- and zwitterion-F^- interactions granted these membranes with the ultrahigh Cl^-/ F^- permselectivity (P_{Cl^-}/ P_{F^-} = 24), enabling high fluoride retention and high chloride passage even from saline mixtures of NaCl and NaF.

29c - Membrane Bonding by Capillary Filling with Viscous Polymers: Infiltration Kinetics and Bonding Strength

Jaylene Martinez (CU Boulder Mechanical Engineering)* ; Masoud Aghajani (Membrane Science, Engineering, and Technology (MAST) Center, Department of Mechanical Engineering, University of Colorado Boulder); Yinan Lu (University of Colorado Boulder); Adrienn

Capillary filling of a porous membrane by a polymer melt is important for membrane device manufacturing. However, a detailed study on pore-filling kinetics and resultant bonding strength have not been reported in literature. In this work, we measure the kinetics of capillary filling a viscous isotactic polypropylene (PP) in microporous polyethersulfone (PES) membranes with varying degree of hydrophilicity. The time-dependent infiltration depth, the microscopic details of the bonded interface, and the contact angles of PP on the PES surfaces were experimentally determined. The results show that the capillary filling kinetics of the membranes at 180 oC appear to be better described by models such as the Cai model that incorporate membrane pore structures, compared with the basic Lucas Washburn model suitable for isolated cylindrical pores. The infiltration kinetics at 200 oC appear to be significantly slower than predictions of both models, which is attributed to the pore deformation/collapse due to the capillary pressure at 200 oC. The resulting bonding toughness, quantified by a modified T-peel test, appear to be dominated by the fracture toughness of the membranes and weakly decreases with the increase of infiltration depth.

29d - Connecting Solute Diffusion to Pore Morphology in Self-Assembled Triblock Copolymer Membranes

Anthony J Cooper (University of California, Santa Barbara)* ; Michael Howard (University of Auburn); Kris Delaney (UC Santa Barbara); Sanket Kadulkar (University of Texas at Austin); Glenn Fredrickson (UC Santa Barbara); Venkat Ganesan (University of Texas

Block copolymers self-assemble a variety of morphologies useful as porous water-treatment membranes. A key challenge is to determine which morphologies maximize the flux of water while selectively rejecting contaminants. Here, we generate equilibrium and nonequilibrium ABC triblock copolymer morphologies using self-consistent field theory (SCFT) and use them in a kinetic Monte Carlo (kMC) model for solute diffusion. Our model excludes transport through the A-block membrane matrix, confining it to the B-block brush that expands when the sacrificial C-block is removed to establish a pore network. These effects are embedded in the SCFT and kMC workflow to examine the roles of mesoscale morphology and internal pore structure in determining solute diffusion rates in porous block copolymer membranes.

30a - Coronavirus and Bacteriophage Removal Mechanism in Membrane Distillation

Mukta Hardikar (The University of Arizona)* ; Luisa Ikner (The University of Arizona); Varinia V Felix (The University of Arizona); Luke Presson (University of Arizona); Andrew Rabe (The University of Arizona); Kerri Hickenbottom (The University of Arizona

The persistence of pathogenic microorganisms in treated wastewater makes disinfection a crucial deliverable in wastewater reuse technology to ensure public health safety. In this study, membrane distillation (MD) is explored as a treatment process for wastewater reuse and virus removal. MD is proposed to provide a dual barrier for virus removal as the high feed temperature can reduce virus concentration and the membrane provides a physical barrier by permeating only the vapor phase. Membrane processes such as ultrafiltration and reverse osmosis (RO) have been shown to achieve high virus and other contaminant removal for wastewater reuse application. However, RO also produces a concentrate stream which contains high concentrations of pathogens and contaminants that often requires treatment and volume reduction before disposal. MD can potentially treat RO concentrate to produce distillate that would augment potable water supply and reduce volume of concentrate to be disposed. However, the rejection of pathogens in MD needs to be quantified for application in water reuse.

In this study, virus removal was tested using a bench-scale MD system with RO concentrate as feed spiked with two non-enveloped phages (MS2 and PhiX174) and an enveloped pathogenic virus (HCoV 229E). The effect of temperature and membrane rejection was quantified in separate tests for the three viruses. At typical MD operating temperatures (greater than 65°C), viable concentrations of all three viruses were reduced by thermal inactivation by more than 6-log10 for MS2 and PhiX174, and more than 3-log10 for HCoV 229E. Membrane rejection was measured to be greater than 6-log10 for MS2 and PhiX174, and greater than 3-log10 for HCoV 229E. MD therefore provides a dual barrier for virus removal and a fail-safe mechanism in which viruses are thermally inactivated before water vapor permeates the membrane, thus adding a supplementary barrier to virus removal in case of membrane failure or wetting.

30b - Elucidating the inherent fouling tolerance of membrane contactors in ammonia recovery from wastewater (V)

Abhishek Dutta (University of British Columbia); Jongho Lee (University of British Columbia)* ; Sifat Kalam (University of British Columbia)

The recovery of ammonia (NH3) from wastewater has garnered interest due to its potential as a fertilizer, refrigerant, and hydrogen carrier. Membrane contactors (MCs) are an effective process for NH3 recovery from diverse wastewater streams, but the impact of fouling on the process has seldom been explored. Here, we examine the dependency of membrane fouling on process conditions and elucidate the inherent system tolerance to fouling in NH3 recovery using MCs. In our isothermal MC system, a hydrophobic PVDF membrane was employed between NH3-rich anaerobically digested food wastewater at pH 10.2 (feed) and sulfuric acid stream (draw). While demonstrating >90% recovery of NH3, our experiments indicated minimal membrane fouling despite the high organic foulant loading in the feed. We then performed a membrane distillation (MD) experiment to investigate the impact of transmembrane water vapor flows and feed pH on fouling. The membranes from MD tests were severely fouled, suggesting the acceleration of fouling in presence of water vapor flows across the membrane. Despite the severe fouling, when these membranes were employed back in the isothermal MC, >70% NH3 recovery from the wastewater was observed. This markedly high NH3 recovery from the fouled membranes was explained using our mass transfer model. Our study shows that (1) isothermal MC processes possess an inherent tolerance to membrane fouling due to the absence of transmembrane water vapor flows and high feed pH; and (2) even in the event of severe membrane fouling, high NH3 recovery is attainable, showing excellent applicability of the MC process for NH3 extraction from diverse wastewaters of high fouling potential.

30c - Evaluation of Direct Heated Vacuum Membrane Distillation Process using Module-scale Simulation

Yiming Liu (UCLA)* ; Jingbo Wang (University of California, Los Angeles); Unnati Rao (UCLA); Bongyeon Jung (UCLA); Erfan Sedighi (UCLA); Tzahi Cath (Colorado School of Mines); Nils Tilton (Colorado School of Mines); Craig S Turchi (NREL); Sungtaek Ju (UCLA

Membrane distillation (MD) is an emerging thermal desalination technology capable of treating high-salinity waste brines. Standard MD processes rely on the water itself as the thermal energy carrier. An alternative approach is to provide heat directly to the membrane surface through the use of composite membrane materials that incorporate thermal energy carriers, which funnel externally provided heat to the membrane/water interface or into the feed channel itself. In our system, a metal shim and a metal mesh are used as thermal energy carrier to heat the membrane and bulk feed stream, respectively. By providing heat to where it is needed, this direct heat delivery has been proven to enhance the thermal efficiency of the MD process as well as increase the potential for single-pass water recovery. In this study, we develop a module-scale process model that captures temperature, salinity, water transport and energy flows during the steady-state operation of standard vacuum MD (VMD) and direct-heated vacuum MD (DHVMD). We explore how shim heating and mesh heating impact both temperature polarization and concentration polarization along the module. Specifically, we model vapor and heat fluxes and compare the performance of the standard VMD and DHVMD configurations. The model shows that direct heat input either to the bulk feed or the membrane surface under realistic conditions increases single-pass water recovery by a factor of 4, and lowers specific thermal energy consumption by ~25%. In addition, we examine the effect of operating conditions on system performance of both DHVMD and standard VMD. Furthermore, we discuss possible designs for a large scale DHVMD that incorporate heat recovery systems, and achieve high water recovery and high gained output ratio (GOR). Overall, these results demonstrate that DHVMD can achieve a higher desalination performance than standard VMD, and has the potential of scaling up from a bench-scale system.

30d - Computational fluid and thermodynamics simulation for direct contact membrane distillation using hollow fibers: scalable meshing and decoupled heat transfer

Albert Kim (U. of Hawai'i)*; Hyeon-Ju Kim (Korea Research Institute of Ships and Ocean Engineering); Deok-Soo Moon (Korea Research Institute of Ships and Ocean Engineering)

Membrane distillation studies often require rigorous simulations of coupled momentum, mass, and heat transfer phenomena. Hollow fiber modules are preferred in industrial applications due to their high packing ratio, resulting in many fibers packed in a vessel. In hollow fiber membrane distillation processes, computational fluid dynamics (CFD) simulations of multi-physics require high-quality meshes and decoupling algorithms for efficient and accurate calculations. Due to the distinct geometric characteristics of HF packing structures, a scalable meshing method is of great necessity but has not been actively researched. Besides, intrinsic coupling of convective and conductive heat transfer phenomena should be overcome to provide reliable modeling results, in a reasonable amount of computing time. This work developed (1) a numerical method to generate hexagonally packed structures of many fibers by forming a hexagonal unit-cell, consisting of the lumen, membrane, and shell regions; and (2) a novel decoupling algorithm of the heat transfer to avoid inaccurate approximations. Theoretical methods and numerical algorithms developed in this study (Journal of Computational Physics 427 (2021) 110042) can contribute to the improved scalability of CFD simulations from lab-scale modules to pilot-scale systems.

AUTHOR INDEX

North American Membrane Society (NAMS)
University of Arkansas
Ralph E. Martin Department of Chemical Engineering
3202 Bell Engineering Center
Fayetteville, AR 72701-1201
USA

ISBN 978-1-7138-3531-8

CURRAN ASSOCIATES INC.
proceedings.com